# Exterior Finishing

**Fine Homebuilding**

# BUILDER'S LIBRARY

# Exterior Finishing

## Siding, Roofs, Decks and Porches

The Taunton Press

Library of Congress Cataloging-in-Publication Data

Exterior finishing: siding, roofs, decks and porches.
       p.     cm. — (Builder's library.)
    At head of title: Fine homebuilding.
    "A Fine homebuilding book" — T.p. verso
    Includes index.
    ISBN 1-56158-065-1 (hardcover)
    1. Roofs — Miscellanea.   2. Exterior walls — Miscellanea.   3. Decks
(Architecture, Domestic) — Miscellanea.   4. Porches— Miscellanea.
I. Fine homebuilding.   II. Series.
TH2401. E943   1993                                     93-10981
690'.83 — dc20                                       CIP

**Taunton**
BOOKS & VIDEOS

*for fellow enthusiasts*

Cover photo: Kevin Ireton

First printing: June 1993

Printed in the United States of America.

A FINE HOMEBUILDING Book

FINE HOMEBUILDING® is a trademark of The Taunton Press, Inc.,
registered in the U.S. Patent and Trademark Office.

The Taunton Press, Inc.
63 South Main Street
Box 5506
Newtown, Connecticut
06470-5506

# C O N T E N T S

# I N T R O D U C T I O N

T he exterior of the house presents major questions of form and function. While protecting the interior and determining how the outside is used, the exterior is also the visible face of a house, the part that neighbors, visitors and passers-by first notice and probably remember the house by. So, important decisions have to be made on materials and style, and workmanship must be superior. Anything less jeopardizes the integrity of the structure as well as its appearance and value.

The 25 articles in this book provide practical ideas and guidance from experienced craftsmen. From roofing with slate, tile, cedar or asphalt shingles to siding with clapboards, shingles or board-and-batten; from replacing rain gutters to building decks and repairing porches, these articles will expand the repertoire and advance the skill of professionals and amateurs alike.

*—Jon Miller, Associate Publisher*

A footnote with each article tells you when it was originally published. Product availability, suppliers' addresses and prices may have changed since the article first appeared.

# Trimming the Front Door

Ripped, mitered and reassembled interior moldings
add a custom look to a front entry

by Richard Taub

***Inspired entry.*** Nearly all the trim—pilasters, mullions, corner blocks and crown—comes from leftover
interior poplar casing. The field to which the trim is applied and the cap on top are clear-pine stock.

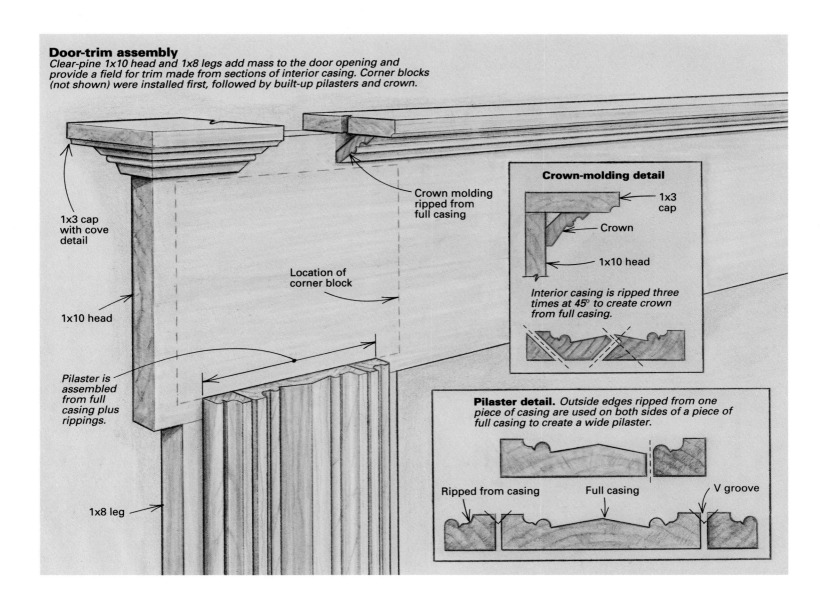

**Door-trim assembly**
Clear-pine 1x10 head and 1x8 legs add mass to the door opening and provide a field for trim made from sections of interior casing. Corner blocks (not shown) were installed first, followed by built-up pilasters and crown.

1x3 cap with cove detail

1x10 head

Pilaster is assembled from full casing plus rippings.

1x8 leg

Location of corner block

Crown molding ripped from full casing

**Crown-molding detail**

1x3 cap

Crown

1x10 head

Interior casing is ripped three times at 45° to create crown from full casing.

**Pilaster detail.** Outside edges ripped from one piece of casing are used on both sides of a piece of full casing to create a wide pilaster.

Ripped from casing    Full casing    V groove

---

Consider the front door, if you will. There is something about this entrance that, if done tastefully, really invites a person into the home, eager to see more. Why, then, are so many front doors stamped in standard, prefabricated trim? You know what I mean: the same old dentils, sunbursts and broken pediments. They're usually southern yellow pine and can cost anywhere from $200 to $350. My partner, Steve Vichinsky, and I recently built a Victorian house with a large hip roof, an octagonal turret and a wraparound front porch. Considering all the work we had put into this house, I wanted to do something special with the front door.

My goal was to build something quickly that was pleasing to the eye and well made. I took a spin around town, looking at the ornate, older homes here in Amherst, Massachusetts, to get design ideas. A day and a half later I had a front door with custom trim (photo facing page) that looked great and was less expensive than the prefab stuff. Plus it was a lot of fun to make.

**Interior trim used outside**—A local millwork company had milled up all my interior trim for the windows and the doors (photo right). The casing was ¾-in. by 3¾-in. poplar. The first ⅜ in. on both sides of the face was milled square to the edges (detail drawing above). Next to these flat sections were 3⁄16-in. coves and then a ⅛-in. V-groove. Two 5⁄16-in. beads were next to the V-grooves. The rest of the trim came to a beveled peak in the center of the casing. I also had plinth blocks made for the doors, bullnose pieces to separate the casing from the plinth blocks, rosettes for above the doors and the windows, and detailed baseboard.

As I stood looking at a finished, trimmed-out room, it suddenly hit me: "Why can't I use the interior trim outside?" I stepped outside. Thinking of the houses I had seen, I knew this front door had to be built out in both width and depth to match the scale of the house. The interior trim alone wasn't beefy enough.

**Fattening the trim**—To give the opening more width, I nailed 1x8 clear-pine stock to the sides of the door and 1x10 over the top (drawing above). The head, or top piece, extends 1 in. past each leg. This gives a nice effect where the legs

**Interior detail.** Custom-milled corner blocks and casing inside the house inspired the design for the front-door exterior trim.

Drawings: Bob Goodfellow

From *Fine Homebuilding* (February 1993) 79:46-49

**Cutting the corner-block components.** Two 45° cuts make a triangle, and four of these triangles make a square corner block. The full width of the casing was used for the end corner blocks.

**Another use for the biscuit joiner.** After carefully cutting and aligning the four triangles, biscuits are inserted in slots to keep the joints tight through all of New England's weather cycles.

**A closer look.** Both the large corner blocks and the small diamonds were assembled with biscuits and epoxy (above), then nailed to the pine. Note how the overhanging head aligns with the siding (left).

meet the head, later accented by the siding. An overhanging head is a common feature of most fancy, exterior door trim. I now had a blank field of pine with which to work.

I tacked a short length of the 3¾-in. interior casing to a vertical 1x8. It looked good but was too narrow. I took another piece of casing and ripped on the table saw a 1-in. strip from each side. I added these strips, which included the ⅜-in. flat, the ³⁄₁₆-in. cove and the ⁵⁄₁₆-in. bead, to both sides of the first piece of casing. With a 1-in. strip on each side, the flat section facing in, the resulting ¾-in. flats felt heavy. So I set my table-saw blade at 45° and put a light chamfer on the edges of the rips and the casing. When I put the three pieces together again, the resulting V-grooves broke up the ¾-in. flats, and the width of the three pieces together looked good. All I needed was a full-length version for each side of the door.

**Mitered corner blocks—**Now I had to come up with something for the head. I thought of using the interior rosette corner blocks, but they were too small. Then I realized that I could make my own corner blocks, once again using the interior trim. I did some tests by putting together mitered sections of the casing with hot-melt glue (see sidebar facing page). Cutting the mitered pieces really isn't too tricky, but if you don't cut the pieces accurately, none of the profile lines will meet. With a sharp blade on my chopsaw, I cut the the casing across the face at 45° and again at the opposite 45° to make a right-triangle offcut (top photo, left). Four of these triangles together made a corner block. I tried different sizes and settled on the largest one, which had triangles cut from the full width of the casing. Large corner blocks seemed more dramatic and better suited to the house. I particularly liked the border that the ⅜-in. flat created around the perimeter of the corner block.

Now that I had my corner blocks, the trick was to join the four pieces permanently. Having just finished joining all the interior trim with my biscuit joiner, I realized that joining the mitered corner blocks with biscuits would be a perfect application for this tool (middle photo, left). I numbered the triangles and put them together face-down with the joints aligned. I marked one line across the center of each joint, cut the biscuit slots and then glued the corner block together (bottom right photo, left) with highly weather-resistant epoxy.

**Laying out the field—**Twelve hours later, when the epoxy had dried, I centered the two corner blocks over the 1x8 legs and nailed them in place. Then I measured the height of each corner block and cut two lengths of casing and four lengths of strips to fit tightly beneath the corner blocks. I centered and nailed a piece of casing on each 1x8 leg. Then I toenailed the smaller rippings on each side of the casing to draw them in tight and countersunk all the nails. Now I had two 5¾-in. pilasters on top of my 1x8 stock, leaving a ¾-in. reveal on each side. This gave me the depth that I wanted. The depth showed up more when the trim was painted two different colors (for more about color schemes, see *FHB* #74, pp.

## Using hot-melt glue

When I was in the Wendell Castle Workshop (a furniture-making school outside Rochester, New York), I found, like most other woodworkers, that there is nothing like mocking up a piece, either full scale or scaled down, that you've worked out on paper. A mock-up quickly tells you whether or not the piece works technically and aesthetically.

**Mock-ups and miter clamps—**For mock-ups, hot-melt glue is incomparable. Once that gun gets hot, the glue melts out and sets quickly (60 seconds to 90 seconds). You squeeze out a bead of glue, stick your pieces together and move right along. Before long, you have a working model in front of you. The only problem with hot-melt glue is that it's so easy it can make you lazy. In situations where you need to see if more intricate joint details will work, such as a mortise-and-tenon joint instead of a bridle joint, put aside the glue gun and do the joint the way it should be done.

When I glue up long mitered joints, joining the sides of a 12-in. box, for example, I make triangular clamping blocks. I lay a bead or two of hot-melt glue on the blocks and stick them on each side of all the miter joints, centered from top to bottom (bottom photo, right). A C-clamp pulls the joint together as it grabs the clamping blocks. This puts clamping pressure directly on the center of the miter joint. It's a good idea to keep the glue gun hot, in case a block comes loose during a glue up. This could be a disaster, but if you're fast, you can remelt the glue, squeeze a bit more on, stick the block back down, let it cure for one minute and continue with the clamp up. The glue is so elastic that it easily pares off with a sharp chisel and can be further cleaned by some sanding.

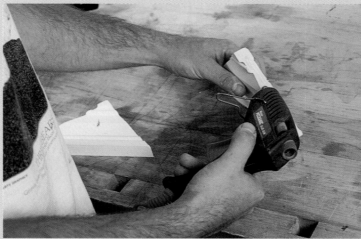

**Melt, aim, shoot.** This trigger-fed gun squeezes a hot bead of glue from its tip. The glue sets quickly as it cools, making it great for sticking together a prototype to see how it'll look.

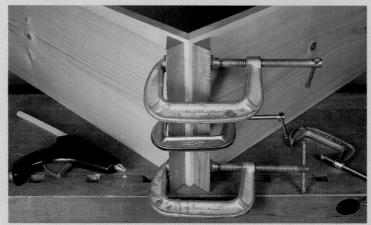

**Temporary clamping blocks.** To clamp miter joints, triangular clamping blocks can be made with hot-melt glue. The blocks can be easily removed later without damaging the mitered pieces.

If you plan to use clamp blocks to pull a piece together, and the piece needs a lot of clamping pressure, blocks glued on with hot-melt glue will never do. Once you start to really torque down on the clamps, the glue bond will break.

I always clamp up the pieces dry before using glue. Dry clamping tells me if the piece will pull together with minimal pressure or if it needs more serious persuasion. Also, I always have another plan ready if the glue fails to hold up under the clamp pressure—maybe some other conventional clamping means, such as band clamps (which happen to work well on mitered frame pieces of any dimension).

**Glues and guns—**Even though hot-melt glue is fairly strong, it cannot take the place of other wood glues. I don't recommend hot melt for gluing a finished piece because it cures so fast that you can't get a tight joint.

Hot-melt glue just doesn't have the holding strength of yellow glue or epoxy.

As for glue guns, there are many on the market, and they are all fairly comparable in both design and price (about $10 to $25). Some, like my Master Mechanic 208-MM (top photo, left), are trigger-fed. This model loads easily; just slip the stick of glue in the back, and the trigger feeds it through the heating element. My gun gets extremely hot—about 380° F at the nozzle tip—so the glue squeezes out very nicely. I've found that the less-expensive models (usually not trigger-fed) don't melt the glue as nicely and won't allow those few extra seconds to work with the piece.

Hot-melt glue itself comes in an array of types and colors. The yellow sticks are basic, all-purpose hot-melt glue and are good for most household projects and repairs. I use the clear hot-melt glue, which is made for craft and hobby use, as well as for building or repairing stuff that doesn't take a lot of pressure or abuse. I wouldn't use it to repair a dining-room chair, but I would use it to fix a loose piece of trim. Hot-melt glue will bond to most porous materials within 90 seconds. There is also a brown hot-melt glue used as a fast-setting waterproof caulk for filling seams. I do not use this glue. Waterproof siliconized acrylic caulk gives you a lot of working time to fill a seam, and it cleans up easily. To get the same results with brown hot-melt glue, you must squeeze out the glue quickly and subject yourself to the heat of the glue as you press it into the seam with your finger. In other words, you would never get a good, clean, painless job. However, the brown hot-melt glue is useful for gluing something that might be exposed to moisture. —R. T.

---

40-45). The mullions between the sidelights and the door are much narrower, so I used the center beveled section of the casing ripped to 1½ in.

I now had an empty field of 1x10 trim above the door and the sidelights. Because the house has many angles and shapes, I thought diamonds might look nice in this empty field. So once again I started playing with mitered sections of the casing and came up with a smaller corner block. I made three identical corner blocks biscuited and glued together just like the large corner

blocks. Turning them on their pointed ends gave me the diamond effect I was looking for. I nailed them onto the 1x10 head trim (bottom left photo, facing page), one above the center of the door and one over each mullion. All nails were countersunk to minimize stains; the painters would later fill these holes before painting.

**Can you top this?—**Now I needed to cap the head jamb. I routed a cove detail on a piece of 1x3 clear pine and nailed it, cove-side down, to

the head. It made a little shelf. Then on the table saw I ripped another section of casing, making three passes with the blade at 45° to create a nice crown-molding detail (drawing, p. 9). I cut 45° compound miters on each end, centered this piece beneath the 1x3 and returned the crown into the sheathing to complete the trim. □

---

*Richard Taub is a home-builder and furniture-maker in Amherst, Mass. Photos by Rich Ziegner except where noted.*

# Painting Exteriors
## Surface preparation is the name of the game

by Robert Dufort

It happens time and again. A simultaneous cheer and sigh of relief go up in the neighborhood when the funkiest house on the block finally gets a new paint job. And then within a couple of years, the seams start popping, the paint starts peeling and ugly stains run down the façade: the face-lift has fallen.

It doesn't have to be that way, but in order to forestall the effects of sun and moisture on painted exterior wood surfaces, we need to first understand the difference between jobs that succeed and those that fail. In this article I'll talk about the products and techniques we use to prepare and restore wood siding and trim as a prelude to repainting. The photos here document a job we did on a redwood Queen Anne Victorian house located in the Pacific Heights district of San Francisco (photo right), but most of the principles involved also apply to newer homes.

**Extensive preparation**—An older home in need of a paint job often requires a good bit of work first. What we call a "painting restoration" involves not only cleaning, sanding and painting the building, but also removing rusted nails and resecuring loose trim; replacing decayed moldings; making windows operational; treating non-structural dry rot with epoxy consolidants; inspecting gutters, downspouts, flashings, and chimney caps for rust; and scrutinizing all exterior surfaces to establish any possible areas of future concern. Although this Queen Anne had been given a face-lift eight years before that included extensive paint removal, patching and repairs, the work had not held up very well. Why?

Believe it or not, the first culprit was the use of oil-base paints, both for primer and some of the finish coatings (more on this subject later). Second, the pervasive dry rot was not effectively treated. Many of the decay cavities

Almost a century old, this Queen Anne Victorian house is decorated with redwood trim and plaster garlands. A job of this scope begins with a thorough cleaning done from a scaffold that allows access to every exposed portion of the building.

had been patched with polyester resin autobody putty, which failed quickly. Third, many areas that should have been stripped of old paint were not—typically a function of budgetary constraints.

**To scrape or to torch?**—There are three primary causes for paint to fail, and they can occur singly or in combination.

First, most paint failures are caused by loss of flexibility of the paint film. Wood buildings are subject to enormous stresses from heat, cold, humidity changes, rain and ultraviolet light. In response to these stresses, the wood expands and contracts. Most older buildings are covered with several layers of oil-base paint, which is relatively inelastic. The wood flexes,

the paint film does not, and surprise—the paint peels, cracks, crazes or craters.

Second, paint applied to a dirty, chalky or glossy substrate can't get enough of a grip on the building to last very long. And third, paint over wood that has moisture trapped inside it will soon pull away from the wall. Other culprits are rusting nails and sheet metal, loose moldings, caulk and putty failures, splitting wood, dry rot and salt accumulations. None of these conditions is unusual to find on an aged Victorian façade. In fact, we are pleasantly surprised when we do not find them.

A good manual that can help you to identify types and causes of paint failure is *The Paint Problem Solver* (Painting and Decorating Contractors of America, 3913 Old Lee Highway, Suite 33-B, Fairfax, Va. 22030; 703-359-0826).

**Analysis**—All paint jobs should begin with an analysis of the existing conditions. From the symptoms of paint failure we determine the underlying causes and then decide upon the proper cure. Some areas may require total paint-film removal, and the method of removal will have to be decided—torching, heat gun or belt sanding (we've experimented with chemical paint removers, but so far we haven't found any that are as effective as heat removal). Other areas may need only a light scraping and hand-sanding. These decisions are made by balancing the solution to a problem against the budget. For example, while in many cases it may be preferable to completely strip a given surface (expensive and time-consuming), it is sometimes possible to perform an adequate job by scraping and sanding. How do we decide whether to scrape and sand or to torch? Like many things in the trades, the right call is a combination of past experience and intuition. If half the paint on the siding of a building that faces the weather is alligatored, bubbled and crazed, then it is a safe bet to assume that the other half is questionable and should be removed as well. If there are a few areas of peeling paint caused by dry rot or rusty nails, it's probably okay to leave the paint film intact. It is also less likely that paint on sheltered portions of the house will need complete removal.

Our analysis of the Queen Anne was pretty straightforward. On the northern corner, alligatoring and peeling paint resulting from heavy accumulations of paint and lax preparation needed to be stripped off. On the south and east street sides, in addition to symptoms similar to the rear, most of the peeling was on the siding and bay windows—areas that had been torched last time. There were several causes. For one, untreated dry rot, especially around rusted nails, caused the paint to fail because "punky" wood is not a good substrate for paint. Second, patches made with auto-body filler failed because of its relative rigidity. As the wood expands, the patch does not, causing the paint to crack. Then water enters the crack, gets behind the paint and causes it to peel (while further feeding the dry rot). Another cause of the premature paint failure was the use of oil-base primer and trim finish. This may come as a surprise to some, but acrylic-latex technology has progressed to the point that one can now say that the use of oil-base paints actually contributes to the premature failure of many exterior wood paint jobs. I feel like I've just started an argument, so I had best try to support this near heretical position.

**Oil versus acrylic latex**—On this Queen Anne many of the surfaces that had been stripped last time were nonetheless checked and blistered. This was caused primarily by hairline cracks in the wood. When wood ages, particularly on exteriors, it dries out a little, deforms, bends, gets hairline checks and splits, and sometimes even pops nails out. If the paint film can't flex with the wood, it has to crack too, and if the peeling paint is cracking along the line of the wood grain, you can bet that the paint is brittle. Heavy accumulations of old paint crack and peel for the same reasons, but the cracks often do not follow the wood grain—instead they have a "crazed" or "alligatored" look with lines running in a reptilian scale pattern. In either case, removal is the only long-term solution.

We knew that the building had been primed and partly finish-coated with oil-base paints. We also know from research literature that oil-base paints are relatively inelastic. Our conclusion: oil-base primers and finishes are to be avoided whenever possible on exterior wood surfaces because they simply do not move with the wood.

I realize that my conclusion is not going to be readily accepted by a large group of both casual and professional painters, and I agree that on surfaces subject to direct abrasion and cleaning, such as handrails, cabinetry and kitchen walls, oil-base paints will often outlast their acrylic counterparts. I will also admit that oil

*A torcher at work.* The quickest way to remove failing paint from wood siding is to scrape it off after softening the paint with a propane torch. When doing this work, it's imperative to wear safety gear and to have fire-fighting equipment nearby.

primers seem to do a better job of displacing chalk, especially on houses that have been painted with marine paints (which are designed to slough off). But given a soundly prepared surface, an acrylic resin binder provides superb adhesion, as well as the needed flexibility to move along with the wood without cracking. Not only has my personal observation convinced me of this, but there is also a large and growing body of research work and experi-

ments by organizations such as the American Plywood Association and the Forest Products Laboratory that verify these conclusions.

**Scaffolding**—Although ladders or rope-and-pulley "falls" may seem less expensive initially than renting pipe scaffold, they are simply not as safe, nor as easy to use. Most jobs can be done faster from pipe scaffolding, especially those requiring aggressive surface preparation

Photo: Staff

From *Fine Homebuilding* (August 1990) 62:36-41

**Failure.** Once paint has started to blister, you can be sure that the bond between the first layer of paint or primer and the wood has failed. When this happens, it's best to take the surface back to bare wood.

A heat plate is an electric paint remover that softens paint without an open flame. This method avoids the hazards of burning paint scrapings falling around the base of the building.

and repairs. Purchase and storage of pipe scaffolding, especially in the amounts required for our larger jobs, is simply uneconomical. In addition, the ability to safely erect scaffold is a trade unto itself, and should not be underestimated; we sub it out. For the last 11 years we've used City Scaffold to erect all of our jobs.

The initial cost for scaffolding on the three sides of this building was $2,595 for the first month. A second month's rental, necessary because of the amount of prep work and repairs, was charged at 30% of the base fee—$780.

**Washdown**—The first step on every job is to clean the building to remove accumulated dirt, mildew, fungi, chalking and rain salts—pollution residues that are borne on the backs of raindrops. Power washing with a compressor is the preferred method, particularly on larger jobs. These compressors can be rented, usually for about $75 per day. A garden hose is attached to the compressor, which in turn sends the pressurized water through another hose and out a hand-held "gun." The gun has several tips that vary the spray pattern, or fan, from 0° (a fine, very intense spray) to a wide 60°. Only the widest tips should be used on wood buildings. The power washer can develop pressures up to 3,500 psi, so it is very important to avoid bodily harm: be careful not to shoot yourself at close range with the pressurized water.

If the building has a lot of mildew and fungi, it's a good idea to first spray the affected areas with a 5 to 1 solution of water and household bleach. A "chemical injector" attachment can sometimes be rented with the power washer to add the bleach. Detergent can also help. We use one called Simple Green (Sunshine Makers, Inc., Huntingdon Harbor, Calif. 92649; 714-840-1319) that can be applied with the power washer. We don't use TSP (trisodium phosphate) any longer—we've found it to be overkill for our purposes and harmful to landscaping. Additional hand-scrubbing may be required for heavily soiled areas. Even though the power washer can remove some peeling paint, remember that its function is to wash, and that you can damage your building by forcing water into the wood. Also, somebody needs to keep an eye on the windows from the inside. They will often leak, even though we avoid spraying them directly.

A towel at the bottom of each window keeps leaking water from causing a problem.

**Taking off the old paint**—Many professional painters in San Francisco remove deteriorated paint films on wood buildings, using propane torches with a live flame (bottom photos, facing page). This method is dangerous, and the risk of fire is real. Compounding the hazard is the possibility of lead poisoning as old paints are vaporized. So why torch? Professional painters use this method because it's the fastest way to take off old paint. We discourage amateurs from using flame-producing tools, and suggest instead that they use heat guns or heat plates (good versions of both are available from Master Appliance Corp., 2420 18th St., Racine, Wisc. 53401; 414-633-7791). While safer, these alternatives are still a fire risk and must be used with care and respect.

With any of these methods it is necessary to wear a respirator (not just a safety mask), heavy gloves, long sleeves, and safety goggles. Prior to applying heat, one should lightly hose down the area, aiming especially in cracks and joints, as this is where most fires start. Be-

Photo: Doug Keister

fore moving on to another area, or before taking a break, hose the same area down again.

The heat plate is placed against the surface being stripped (right photo, facing page), and held there from 10 seconds to over 30 seconds, depending on thickness and types of paint being removed. When the paint becomes soft and bubbles a little, you scrape it off.

A heat gun resembles a souped-up hair dryer, though it's a little bigger, heavier duty and much hotter. I think the best brand on the market is the MASTER HG-501 heat gun (also from Master Appliance).

Propane torches come in two basic formats. A small hand-held propane canister with a screw-on nozzle and flame-head is commonly available, and best suited for beginners. The flame stays small, but is effective for most surfaces, particularly moldings. The setup we use is a 5-gal. refillable propane tank, and attached to it is a regulator, 20 feet of hose and a handle that can accept a variety of flame-heads depending on the size and shape of the surface being torched. Rigs like these are typically available through propane refill stations and professional plumbing-supply stores.

The heat gun and propane torch are used in similar fashion. The nozzle is held from 6 in. to 18 in. from the surface being torched, and is moved back and forth—it is rarely held steady in one place. The object is to heat the paint, not burn it up. Scorching the bare wood is to be avoided, and is usually caused by overlapping the flame onto previously stripped areas or by using a flame that is too large or too hot. Usually, we'll heat up a flat area about 6 in. square, though the area is less with intricate moldings. Work always proceeds from the bottom of the house up. The size of the flame and the distance from the work can vary according to weather conditions, and a pro will alter the variables as necessary.

Our painters use shaped scrapers that are pulled instead of pushed (top photo). They are designed specifically for torch work, but they're also ideal for "dry" scraping. The different shapes of the scraper heads fit many moldings, and we often customize them with a file that is also used to keep the blades sharp. We get our tools from American Paint Scrapers (distributed by Bill Majoy, 183 Clearfield Dr., San Francisco, Calif. 94132; 415-664-2440). Regardless of the tool used, some paint residue is usually left on the surface after the first pass. It is lightly heated again and scraped off.

**Safety**—Many precautions need to be taken when stripping by heat, regardless of the technique. Fire regulations require that the contractor be licensed and properly insured for the work, that a garden hose be within reach of the "torcher," and that a watch be maintained for at least one hour after all burning activity has ceased for the day. In San Francisco, the painting contractor has to acquire an annual permit from the fire department, and then has to register each job site with the local fire station. Often the fire marshall will drive by a job site to check on progress, and we welcome the visit.

Painters need a variety of scrapers to correspond with the nooks and crannies that distinguish Victorian detailing.

Using a propane torch, a painter heats the old paint until it blisters (above). Using a scraper, he pulls the softened paint off the wall (below).

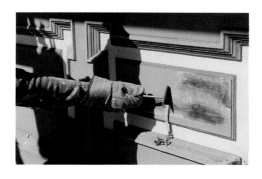

It's worth repeating: most fires are started by a small spark or ember that gets started in a crack or an open seam. Sometimes old rags or newspapers have been stuffed behind boards, and they can ignite easily. That is why we always hose down an area just prior to starting work, and also immediately afterwards. We also keep a crowbar, a hammer and a fire extinguisher nearby in case fire starts in a crack that water can't reach. Paint scrapings usually fall straight down, but they are sometimes still burning so we're careful to keep combustibles such as drop cloths out of the way. We clean up the scrapings with a broom and rake, and bag them for disposal.

**Sanding**—After torching, we sand down all areas with 50- to 80-grit production open-coat paper. We use finer sandpaper (120-grit) around entries and on eye-level surfaces. Any areas that have been scorched need to be sanded down to good wood. Also, glossy surfaces need to be etched or dulled so that the new paint has something to grip—some "tooth." Both of these operations can be done either by hand or with a small electric orbital sand-

er. The orbital sander is a painter's mainstay, and we use ones made by Makita and Porter-Cable. The Makita fits the hand better, but we find the Porter-Cable to be more durable.

Orbital sanders are relatively light, can fit into all but the smallest flat spots and can even handle most larger trim work. We use them to sand down imperfections and paint residue, and to soften the line where stripped wood meets a line of paint that is good enough to leave on the wall. Orbital sanders are not, however, very effective at removing loose or deteriorated paint.

An electric belt sander can be used for removing loose paint from large flat surfaces like siding. We use 3-in. by 21-in. belt sanders, fitted with dust bags, from Makita and Hitachi. These machines are very powerful and fairly heavy, and must be used with care. A fatigued operator can easily do surface damage to the work.

I don't recommend circular sanders and rotary-drill attachments because they are difficult even for professionals to control. We have tried grinders specifically designed for paint removal. They definitely remove paint, but surface damage is unavoidable with these tools, and I think they are best suited for rust removal on metal surfaces.

Regardless of what kind of sander you use, protect yourself against potential poisoning from old lead paints. One study in which our company participated suggested that higher levels of exposure to lead resulted from sanding rather than from torching. Wear a respirator, long-sleeve shirt and gloves while sanding. Change your clothes before heading home and be sure to wash up thoroughly before eating.

**Under the old skin**—On this project we had to get rid of virtually all of the paint exposed to the weather on the front (south) side of the house to uncover the problems causing the paint failure. Remember, much of the paint in these areas had been stripped at the time of the last paint job. These surfaces included siding and ornate trimwork, requiring us to use every scraper in our bags.

As expected, our torching revealed extensive dry-rot cavities, some of which had been patched in the last go-around. None of the patches had held up. On the front and east side, we stripped the lower panel band (much of which was new redwood installed during the last restoration), window sills and the bottom rails of window sashes.

In areas where 80% to 90% of the paint was sound, we decided not to remove the entire paint film. Instead, we used our putty knives and paint scrapers (without the torch) along with sandpaper to remove loose paint. Then we used the orbital sanders to feather the edges of the scraped surfaces.

At this stage we had a building that was clean, with deteriorated paint removed, glossy surfaces etched, imperfections sanded and the mess cleaned up. You might think it was about time to get some paint on the walls. Well, not yet. Next, we inspected the building, looking for the paint-related problems that inevitably show up. This includes a check of loose and

missing trim, sheet-metal flashings, gutters, chimney caps, window putty, rust spots, dry rot, the condition of any plaster details and the condition of the roof. Looking from the inside can often uncover valuable clues to roof problems. Leaks, stains and crumbling wall plaster all hint at defects that will also affect the integrity of the exterior paint job. These fixes are out of our realm, so we call in the right expert to correct the problems. On this job we were fortunate that Winans Construction Company was already in the building doing a kitchen remodel—we only had to holler whenever we came across a questionable condition. Contrary to popular opinion, painters and general contractors can actually work side by side in near harmony.

Loose trim should be resecured, and missing pieces replaced. Care should be taken in renailing, as older wood trim is often very fragile and will split easily unless predrilled. Often, it is best to remove a complicated trim section, carefully take it apart and strip and repair it before reattaching it as a unit. Less complex trim pieces are often easier and less expensive to replace than to repair. Our torching uncovered several small sections of siding that were badly split, so replacement made sense.

**Epoxy patching—**Chances are good that a hundred-year-old wood building will have some loose, rusted nails as well as pockets of soft, deteriorated wood. Often the wood is soft enough to stick a pencil into. We refer to this condition as dry rot, even though in some cases it is caused by a chemical reaction between the wood and the rusting nail. The condition can also be found on window sills (top photo) and at the joints between pieces of trim. In any of these situations, the treatment is the same.

The first step is to dig out the loose, punky wood fibers and rusted nails—we usually do this with our scrapers. We don't worry about cutting back to 100% sound wood, because the two-part epoxy consolidant that we use as a primer helps rebuild weaker areas. After excavating the deteriorated fibers and resecuring any trim pieces, small holes are drilled into the surrounding wood to allow the injection of the epoxy. We often use plastic catsup and mustard squeeze bottles to apply the consolidant. On larger, flat surfaces like window sills, we just brush it on. The epoxy consolidant is absorbed by the wood, killing any fungi and encapsulating and stiffening the wood fibers

*Dry-rot repair.* **Areas of dry rot often occur at the intersections of trim pieces, atop window sills and adjacent to rusty nails. To repair them, the painters first remove the punky wood and rusty nails. Then holes are drilled into neighboring areas to allow a clear epoxy consolidant to be injected into the wood (the dark stains in the top photo show where it has been applied). The next step is to fill the cavities with epoxy fillers, which are white in the middle photo. After two coats of primer and a finish layer of paint, the restored window and trim take on a new look.**

to allow acceptance of the epoxy patching material that follows.

The two-component epoxy patching material is mixed thoroughly and then applied with a putty knife to fill the hole completely. Before the epoxy starts to harden (usually only a matter of minutes) we "blade" over it with a putty knife dipped in lacquer thinner. This shapes and smooths the exposed surface somewhat and minimizes the need for subsequent sanding. If the excavated area is very large, we will often fill the hole partly with epoxy, slip in a redwood plug shaped to fit, and then fill the remainder with our patching material. The epoxy fillers shrink very little, if at all, so most holes only have to be filled once.

After it cures (bottom photos, facing page), the epoxy patch can easily be sanded, filed or shaped with edge tools. This stuff produces a nearly permanent repair, as it literally fuses with the surrounding wood and flexes at a similar rate. About the only things we can do wrong are mixing the wrong ratio of the two components together, or mixing them inadequately—in either case the material might not cure. The epoxy consolidants and fillers we use are available by mail order from Abatron, Inc. (33 Center Dr., Gilberts, Il. 60136; 708-426-2200) and Smith & Co. (5100 Channel Ave., Richmond, Calif., 94804; 415-237-6842). These companies produce epoxies specifically designed for use with wood and are glad to send out informative literature to anyone who asks. Other epoxies can sometimes be found in marine-supply stores. For more on the subject, get a copy of "Epoxies for Wood Repairs in Historic Buildings" by Morgan Phillips and Dr. Judith Selwy (Superintendent of Documents, U. S. Government Printing Office, Washington, D. C. 20402; 202-783-3238).

**Priming**—Now that the building has been prepared, we finally get to do some painting. The initial coat of paint, the primer, serves as a foundation. It is intended to ensure good adhesion between old paint, bare wood, and the finish coats of paint. Usually, a full prime coat on all surfaces is recommended for best results. In some cases, however, it may be possible to "spot prime" the areas of concentrated scraping if the remaining surfaces are in good condition and have been painted with latex acrylic within five years.

We have two favorite acrylic primers: Lucite Exterior Latex Wood Primer-#51 (Olympic, PPG Architectural Finishes, Inc., 2233 112th Ave. NE, Bellevue, Wash. 98004; 206-453-1700); and Sinclair Paint's StainLock Acrylic Latex Primer (Sinclair Paint Co., 6100 S. Garfield Ave., Los Angeles, Calif. 90040; 213-888-8888). Both are tenacious primers suitable for new and old work, and both do an excellent job of sealing redwood and cedar tannins. Lucite is substantially more expensive, but it levels out better than the Sinclair primer. These primers also adhere just fine to oil-base finishes already on the building, as long as the old finishes are sound, clean and deglossed. They can even be applied on slightly damp—but not wet or

***Filling wide cracks.*** **Backer rod is a compressible foam rope that can be squeezed into a crack too wide to caulk. Once in place, the remainder of the crack is filled with caulk.**

soaked—surfaces. Two coats of primer are recommended for all stripped and new wood. One coat of primer is generally sufficient on existing painted surfaces in good condition. We strive for good, full thick coverage because a thick flexible film stretches more than a thinner flexible film does. The converse is true for inelastic (oil) films—the thicker it is, the less it flexes.

I do have to confess that there is one oil primer we use frequently—for rust spots. Our choice is called "Rust Destroyer" (Advanced Protective Products, Inc., 187 Warren St., Jersey City, N. J. 07302; 201-435-6166). We've found it to work better than the traditional films that merely cover-up the rust with an inhibitor.

It is a rare job where the primer (and at least one of the finish colors) is not applied initially with an airless spray gun. The reason is simple economics—it is faster. We do, however, always "brush-in" the primer right after it is applied by gun. After spraying the primer onto an area, we go back over it with our brushes, evening it out and getting it into all cracks, seams and hard-to-reach surfaces.

Spraying outdoors does create one big problem—overspray. This nemesis is especially troublesome in windy locales such as San Francisco. As one client (years ago) said to me, "I didn't notice that painting my neighbor's car was in the contract. I hope you're not going to charge me extra for it." Nowadays we drape large burlap canvas sheets on the outside of the scaffold if we think overspray is going to be a problem.

We have never had any problems painting with acrylics in the fog or in direct sunlight, but in each case the drying time changes. With higher temperatures, the primer can get gummy, sticky and not brush out very well. Below 50° F, the acrylic resins will not properly bond with one another as the paint dries.

The second coat of primer, if needed, is usually applied after the caulking, puttying, and repairs have been performed. This not only ensures that the putty and repairs have a prime coat over them (a good idea), but also cleans up the building, making it easier to spot areas that still need attention.

**Caulking and puttying**—We use caulk to fill areas that are subject to movement, particularly cracks, seams, and joints. As a consequence, we need a caulk that can flex without losing its grip. We primarily use VIP Waterproofing System's #5710 (Flood Company, 1213 Barlow Rd., Hudson, Ohio 44236; 216-650-4070). It's made with a 100% acrylic terpolymer resin that allows the caulk to stretch 5¼ times its thickness before breaking. After initial application it is usually smoothed out with a finger, and then lightly wiped with a damp cloth (for more on caulks, see *FHB* #61, pp. 36-42).

Not all seams should be filled with caulk, however, because it's important to leave passages for moisture inside the house to escape. The seam found between each siding board and the one above is a good example of where not to caulk. But any seam that appears to let rain enter the building should be caulked. Generally, we will do extra cosmetic caulking on the front of a house and be concerned only with waterproofing on less visible sides.

Larger seams often need to be filled initially with something other than caulk—we use backer rod or rope caulk. Otherwise, you can shoot a whole tube of caulk into a joint and still not fill anything. Backer rod is pushed into the seam with a putty knife until it is ¼ to ½ in. past the outside edge of what you want caulked (photo above). The difference is then filled with the caulk. The brand we use is available in a variety of diameters from ¼ in. to 1 in. (Sashco Sealant, 3900 E. 68th Ave., Commerce City, Colo. 80022; 303-286-7271).

We use painter's putty to fill the basic nail hole. The rule here at Magic Brush is: dime-sized holes and larger are filled with epoxy; smaller holes are filled with putty. The brand we use, "Lead Free Putty," (Crawford Products Co., 419 S. Park Ave., Montebello, Calif. 90640; 213-721-6429) is made with linseed oil, and it's not very sticky. To apply it, hold a golf-ball size gob of putty in one hand and push it on top of the hole. With the other hand, push a putty knife into the ball; then draw it back over the hole, leaving a flush, nicely filled hole. No excess is left, and no sanding is required.

By the way, when you buy supplies for painting, you generally get what you pay for. For instance, acrylic resins are more durable than vinyls, and they also cost more. Considering that the difference in cost between the best and second-best materials amounts to but a tiny fraction of the overall cost of a job, it's false economy to skimp on your paints and primers.  □

*Robert Dufort is president of Magic Brush Inc., an interior/exterior painting company in San Francisco specializing in painting-restoration work. Photos by the author except where noted.*

***Framed.*** Painting the relief layer a light color emphasizes the outline of the building, and in the case of this south-facing Victorian, extends the life of the paint. This is a complementary scheme with a light-valued pumpkin-orange body and deep-blue sash.

# Understanding Color

## Before you paint, brush up on the facts

### by Jill Pilaroscia

Against the backdrop of the sky, on a canvas the size of a drive-in movie screen, a freshly painted house can become the toast of the street or a head-shaking embarrassment. Given the expense of a good paint job and how long you (or your client) will have to live with it, it's no wonder that picking the colors can be agonizing. After all, few tasks are as public as painting a house. But there are some guidelines that can help you in the quest for a suitable color scheme.

**Color vocabulary**—Each color has three basic characteristics: hue, chroma and value. *Hue* is the formal term for a color family. For example, burgundy and rose are different colors, but they are the same hue: red. *Chroma* refers to the relative strength or weakness of a color. A very bright lemon yellow has more chroma than a pastel yellow. When applied to color, adjectives like saturated, intense, vivid and rich mean that color has a lot of chroma.

Chances are you've visited a paint store and contemplated the racks of paint chips, where thousands of colors are carefully arranged in relation to one another. The most useful sets of chips contain a color and several *values* of that color. Value is defined as the relative lightness or darkness of a color. A light color is called a *tint;* a dark color is called a *shade*. The relative difference in lightness and darkness is measured on a value scale, which ranges from 0 (black) to 10 (white) (Munsell value drawing, p. 20).

The *temperature* of a color is either *warm* or *cool*. Warm colors—red, orange and yellow—advance visually. Cool colors—blue, green and purple—recede visually. Studies done by Faber Birren, the father of color psychology, revealed that warm colors are most appreciated by gregarious, outgoing people, while cool colors appeal to the cerebral, introverted person. *Neutral* colors—grays, taupes and beiges—evoke very little emotional response and appeal to almost everyone.

Two or more colors, or values of the same color, placed next to one another produce a *chord. Related* chords combine colors of differing values, or colors close to one another on the color wheel (drawing, p. 20). *Contrasted* chords include colors that are separated on the color wheel. Chords are typically arranged in palettes that fall into a half-dozen color schemes (drawings, p. 21).

**Color selection**—How do we apply color theory to choosing colors for a house? Let's begin by considering the house's exterior components. A house is typically composed of three visual elements: a *base* that connects it to the earth or the pavement; a *body* or *midsection* with windows; and a *peak* or *roof zone*, which caps the building and sets it apart from the sky. Each of these elements will usually have its own color.

The shape of the house can be emphasized by highlighting its *relief layer* with colors that con-

***Brick base, tile top.*** **A color scheme needs to take into account the given colors of the building. Here, the author built a split-complementary scheme of light-valued peach and grayish blues and greens to unify the deep-red brick base and the greenish-blue tile roof.**

***Earth tones.*** **A neutral palette of bluish-gray, olive-browns and cream makes a stately foil for landscaping. Note how the darker lower body color of the house grounds it and emphasizes the curve of the bay window painted in a lighter value.**

trast with the body. The elements that make up the relief layer include trim pieces, such as corner boards, quoins, fascias, the bands that express the different floor levels of the house, window casings and sills. Larger assemblies, such as bays, balconies, cornices and entryways, are also part of the relief layer.

Smaller details, such as window sashes, railings, awnings and house numbers, are perceived after we've taken in the background and the relief colors. These details can be accented by painting them a different color than the relief layer. Frequently, they are made of iron, brass, baked enamel or fabric.

It's possible to get by with a three-color palette, but I much prefer working with four to six colors. The problem is that it's often hard to arrange three colors in a manner that makes the most of a house's architectural features. For example, when an architect arranges the massing of a house and applies ornament to it, some portions advance and some recede. A three-color palette allows one body color, one relief-layer color and one accent color for the sash and the doors. With a few more colors, I can emphasize the focal points by setting them apart from the rest of the building. Here's a typical distribution: one major body color; one minor body color; one relief-layer color; one sash color; and one or more accent colors for areas such as soffits or friezes.

The base, which in some cases is also the foundation of the house, will often have a predetermined color such as brick, stone, pigmented stucco or unstained shingles (photos previous page). So I choose the body color first, as it encompasses the greatest surface area. The most basic decision about the body color is whether you want your building to stand out from its surroundings or to blend in. Contrasting or warm colors will make it stand out, while cool colors, or colors that approximate the setting, will minimize its presence. Houses that are close together can be made to feel more private by painting them with cool colors.

As a general rule, the body color should be lighter in value than the base to keep the building from looking top-heavy. On houses without a dominant base, this same look of being *grounded* can be achieved with a light body color above a darker one (photos previous page). The body color should be compatible with the predominant colors of the site, which may include the colors of the soil or the vegetation, the roof tiles or the shingles and the color of neighboring buildings. Important points to remember in choosing a body color are: darker colors read heavier in volume than light colors; a roof that is lighter in value than the body can give the roof an unsettling, floating quality. A way to keep the building together as a composition is to separate a light body and a light roof with a dark fascia band.

Don't forget that color intensity amplifies on a large surface in daylight. If you want a blue building, get a blue with some gray in it or prepare to be overwhelmed. Probably the most common horror story a painter hears is the cry of disbelief that accompanies a client's first glimpse of a wall

painted with a color chosen exclusively from a tiny paint chip.

Next I pick a color for the relief layer. Light trim colors emphasize the building outlines and its openings (photo p. 18). They also reflect light and heat, thereby extending the life of the paint film. Grays are chameleon colors that will pick up and harmonize with adjacent colors. Whites lessen the impact of the colors next to them, while black emphasizes the colors it borders.

Historic color schemes tend to be built around dark trim tones, which look very rich but have a tendency to fade in direct light. Dark trim on a dark background flattens a wall surface and minimizes projections. Dark trim on a light background can make detail appear smaller. You can use these principles to downplay awkward trim or poorly proportioned pieces of the building. Conversely, a special detail, such as a finely wrought balcony or a bay window, can be a focal point when painted with a light color. Wrought iron painted in light colors will appear lacy and decorative. Dark wrought iron appears more formal and traditional.

Choosing colors that have contrasting values is just as important as the colors that you select. For example, a house that is painted in light-valued colors will appear washed out. On the other hand, coloring a building with all dark values can make it appear ominous. Here's a general rule of thumb that can help in selecting values: You need at least four full steps between the body and the relief layer to perceive a significant contrast between the two. For example, a body color of value 5 needs a relief layer of value 9 (see sidebar, facing page).

If a multicolored palette is beyond the scope of a project, my advice is to do something conservative and simple. Pick a neutral body color with a value of 4 or 5, paint the trim and the sash an ivory white and let people know where the entry is with a richly colored front door.

**Accent colors—**The eye tires quickly of too many intense colors, but if used judiciously, they can be just the highlight you're looking for on soffits, panel moldings or a frieze. Use them the way they occur in nature, where vivid spots of color on fruit, flowers and tropical birds are contrasted against backgrounds of muted greens, umbers and browns.

Accent colors don't have to be light in value, but they should be full of chroma. For example, accent colors, such as dark bottle green, burgundy and navy blue, work well on window sashes because they have enough contrast to stand apart from the window panes, which read as a dark gray surface. Black, charcoal or umber sash colors will merge with the glass itself, but this tendency can be counteracted with light drapes or shutters. Light-colored sashes work particularly well on southern exposures, where the first area to fade or peel is the window sash.

Simplify the colors around entryways to keep the transition from outside to inside uncluttered. And put the lightest color in your scheme on the entryway ceiling, where it can bounce light around the door (if you have a large porch ceiling, consider tones of sky blue to create a sub-

liminal sense of openness and light). Using a dark, rich color on the doors reinforces the function of the door as a place to come in—color there can set a mood or make a statement.

**Tabletop color tests—**A good way to develop a color scheme for a house is to photograph its important elevations and details. Take both 4x6 color prints and 35mm slides. Mount the prints on white paper to use as site references so that you don't have to keep running back and forth to check the surrounding colors. Project the slides onto sheets of white paper and outline the basic elements and the relief layers of the house with pencil. Then make a number of copies on white paper so that you can fill them in with colored pencils or felt markers to try out various schemes. I like the Prisma Color pencils and markers made by Berol (you can get these from an art-supply store).

Get paint-chip samples from your paint store in the colors that you're considering. The best ones will have four or five values of the same color on them (Benjamin Moore, Sherwin-Williams and Fuller-O'Brien are three national brands that have extensive paint-chip samples in different values). By lining them up next to one another, you can compare the value relationships between the different colors.

It's tough to duplicate paint-chip colors with felt pens or colored pencils unless you've got a huge selection. But you can get a good range of values by first laying down a base of gray pencil. Use the value chart on the facing page as a guide to calculating how light or dark the base of gray pencil should be. Then go over the pencil with a marker. Another approach is to overlay a base of marker color with a white pencil. Try to work in natural light. Remember that you are approximating colors at this point, trying out ideas without opening a paint can or leaving the kitchen table. Develop a couple of promising schemes.

**Consider the light—**Don't forget the quality of the light that falls on the principal façade as you work with your color samples. The light that falls on a north façade is green and cold, which invites the use of rich, saturated warm colors that can have more chroma than façades facing other points of the compass. Also, the lack of direct sun on a north façade means that colors won't fade nearly as fast.

The harsh light of an eastern exposure will wash colors out in the morning hours—strongly contrasted colors will keep them differentiated. It is most difficult to hold colors on southern exposures, so keep the colors from value 5 and lighter to extend their lives. If you choose darker colors for southern exposures, apply them to parts of the house that can be easily reached for repainting (like the doors) or in areas that are shaded most of the time. You can do just about anything you want with a western façade. The warm, pink light that falls on it in the afternoon is like a giant incandescent bulb.

Not all parts of the country receive the same kinds of light, and varying the color schemes accordingly can be a good strategy. For instance, light-valued colors can enliven houses in

From *Fine Homebuilding* (April 1992) 74:41-45

# Around the wheel

The color wheel (bottom drawing) shows the relationships of the warm and the cool colors to one another. The primary colors—red, yellow and blue—are at the points of an equilateral triangle. Around the wheel, the secondary colors—orange, green and purple—fall on opposite sides of the primaries. Colors on opposite sides of the wheel are called *complementary*. For example, orange is the complement of blue. Primary and secondary colors evoke psychological and physiological responses in people.

*Red* is stimulating. It suggests heat, passion and strength. On the biological level, it can increase blood pressure and body temperature. Red is associated with holidays that signify patriotism, religious fervor, sentiment and romance.

*Orange* has warm invigorating qualities. It conveys the living, pulsating life force but is mellowed and less primitive than red.

*Yellow* is cheerful, high spirited and associated with sunlight. It radiates warmth, wisdom, inspiration and intelligence. Ancients wore yellow amulets as protection against disease.

*Green* is nature's color, associated with relaxation and rejuvenation. Green is subdued and psychologically nourishing. It can lower blood pressure and is frequently used in hospital interiors to assist healing.

*Blue* reminds us of the sky and the water. It has a transparent quality, and it inspires peace and introspection. It can also be associated with depression and melancholy, as in the expression "feeling blue."

*Purple* is mysterious and exclusive. Once costly to manufacture, purple was reserved for royalty and was the imperial color of Rome. It is aesthetic, refined and elevating.

People also develop subjective links with colors that can get positively personal. For example, if you were disciplined in a yellow room as a child, you are likely to have a lifelong aversion to yellow. On the other hand, if you had a happy childhood in a gray room, gray will forever be of comfort. The point is that color and memory can become intertwined and shouldn't be overlooked when it comes time to pick a color scheme. — *J. P.*

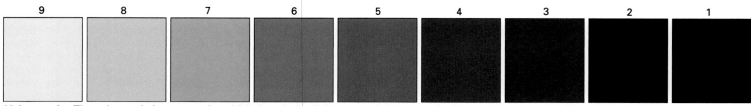

| 9 | 8 | 7 | 6 | 5 | 4 | 3 | 2 | 1 |
|---|---|---|---|---|---|---|---|---|

**Value scale.** The value scale is a means by which the relative lightness and darkness of colors can be compared. It's a black-and-white camera-eye view of a colored image.

(10 = pure white
0 = pure black)

Value scale courtesy of Munsell

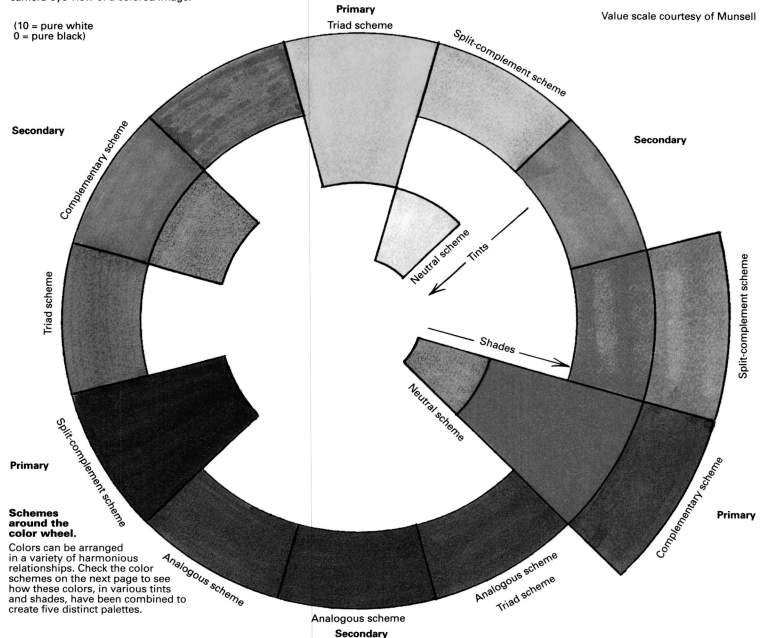

**Schemes around the color wheel.**

Colors can be arranged in a variety of harmonious relationships. Check the color schemes on the next page to see how these colors, in various tints and shades, have been combined to create five distinct palettes.

# APPLYING COLOR SCHEMES

## Related color schemes

**Monochromatic** (photo facing page). Conservative and decorous, a monochromatic palette is one in which areas of the building are painted in different values of the same hue. A classic example is the gray scheme on a Colonial house in which the values range from white to black. A monochromatic brown scheme would range from light buff, midtone café au lait to dark brown. It's pretty hard to go wrong with a monochromatic scheme—the only pitfalls are in letting the values get too close and the chroma too intense. Reds, blues and greens can also be used to concoct successful monochromatic schemes. The photo on the facing page shows one in mossy green that I devised for a shingled Victorian house.

**Neutral.** Slightly more adventurous than the monochromatic approach is the neutral palette, where the dominant surfaces of the building are painted in neutral colors, such as gray, taupe (beige with some gray in it), tan or greige (gray with some beige in it). This reserved background is then enlivened with light-colored trims and rich accent colors on doors and windows. Colonial, Tudor and California bungalow houses often use this kind of palette. Neutral schemes are refined and elegant and go well with landscaping.

**Analogous.** Colors that are close to one another on the color wheel make up analogous palettes—blue, blue-violet and magenta, for example. The accents to enliven this kind of scheme would vary the chroma and the value of the colors bridging blue and magenta. Adventurous Mediterranean style and Victorian houses are often painted in analogous palettes, sometimes with accents in complementary colors for even more zing.

## Contrasted color schemes.

**Complementary.** This scheme is created by using colors opposite one another on the color wheel, such as blue and orange or yellow and purple. A familiar example is the red brick building with green trim. Color schemes based on complements are harmonious and balanced, and studies tell us that this harmony is psychologically calming.

**Split Complement.** Instead of orange, which is the true complement of blue, the two hues adjacent to orange on the color wheel—red orange and yellow orange— are used with blue. The two orange hues work in harmony with each other while the blue adds vibrant contrast. Split-complement schemes work well with Victorian houses, where there are lots of layers and surfaces for color, and with brick or stone buildings.

**Triad.** This palette uses any three colors that are spaced in fairly equal increments around the color wheel. For example, red, yellow and blue make a triad. Moved over one space, red-violet, yellow-orange, blue-green is another triad. Triadic palettes tend to be very showy and are best used on commercial buildings that want to attract attention or on houses occupied by people who want their doorbells rung.  — *J. P.*

Neutral

Analogous

Complementary

Split complement

Triad

**Further reading**
Color theory is a absorbing topic that can be studied from many different angles. Four of the best books that I can recommend are:

• *Victorian Exterior Decoration* by Roger Moss and Gail Winkler, Henry Holt, 118 W. 18th St., New York, N. Y. 10011, 1986. $35, hardbound, 128 pp. A study of 19th century color schemes.
• *Daughters of Painted Ladies* by Michael Larsen and Elizabeth Pomada, Dutton/Penguin USA, P. O. Box 999, Bergenfield, N. J. 07621, 1987. $16.95, softbound, 144 pp. Presents Victorian homes from across the United States.
• *How to Create Your Own Painted Lady* by Michael Larsen and Elizabeth Pomada, Dutton/Penguin USA, P. O. Box 999, Bergenfield, N. J. 07621, 1989. $19.95, softbound, 167 pp. A workbook illustrating the approach of six colorists to creating a palette for the same house.
• *The Elements of Color* by Johannes Itten, Van Nostrand Reinhold, 7625 Empire Dr., Florence, Ky. 41042, 1970. $19.95, hardbound, 96 pp. This excellent book on color theory by a Bauhaus Master of Form contains many plates illustrating the interaction of colors.

The following books are out of print but are worth searching out at your library or second-hand bookshop. *Architectural Color* by Tom Porter (Whitney Library of Design) is a guide to using color on buildings. It has many color plates and a chapter on making color decisions. *Color Psychology and Color Therapy* by Faber Birren (University Books) is a study of the influence of color on human behavior.

For the last word on color standards, measurement and notation, get in touch with Munsell Color. They have everything from charts and texts on hues, values and chroma to a color service that can identify color samples sent to their lab. Munsell Color Products, Macbeth Division, P. O. Box 230, Newburgh, N. Y. 12551; (914) 565-7660.

overcast areas like Seattle, Washington, but beware of too much chroma. Sunny, bright regions can handle saturated colors. For instance, Miami, Florida, is a locale with strong light where houses are often painted in lively, rich colors without looking out of place. The same colors applied to a building in the Midwest, an area with denser light qualities, would look gaudy.

**On the wall—***Never* assume your paint-color choices are correct until you sample them on your building. Even with careful planning, natural light can alter the most thoughtful schemes, turning palettes into discordant disasters.

The portions of wall and trim upon which you plan to test your colors must be cleaned and primed white. Sampling colors over tinted primer or existing colors will never give you an accurate visual test. Get quarts of your chosen colors and make sample patches of the body color in 3-ft. squares. Make sure that all the colors you plan to use touch one another.

You're going to be trying out various tints and shades of the body and trim colors, so have clean mixing pots, stir sticks, rags, cans of black and white paint and a color wheel handy for reference. In addition to changing the value of colors with white or black, you can increase the harmony between colors by adding a small portion of color A to color B. If, for example, you like the body color but the trim is too much of a contrast, add a bit of body paint to the trim paint.

It can take many samples to get a color scheme that works under different lighting, but for an investment as substantial as a paint job, the results are well worth the effort. By the way, paint changes color as it dries. Some get lighter, some darker. So make sure the paint has completely dried before making any final decisions.

When you've settled on a scheme, make samples of the colors by painting them on white cardboard or bristol board. Then take them to a paint center that can do a computer analysis to get the right formula to duplicate the colors. Keep the color samples for your chosen scheme in a safe place for future reference. Put them in a folder so they won't fade.

In order for the colors to look as planned on the house, they need to be applied over a base of white primer. If you expect to paint directly over an existing paint job, be prepared for the old colors under the new paint to affect the outcome.

For exteriors, I recommend semigloss paint because it resists the buildup of dirt that breaks down the paint film. The trade-off is that surface imperfections are more visible. Weathered surfaces should be carefully prepared with scrapers and sanders prior to receiving a coat of white primer (for more information on exterior paint preparation, see the article on pp. 12-17).

Finally, make sure you compare your custom-mixed colors with the paint-chip samples of your final selections. Sometimes they don't match, and it's a lot less trouble to find out before the paint goes on the wall. □

**Monochrome.** This mossy monochromatic scheme was inspired by an acacia tree in front of the house. Note how the dark-valued stairs (about a 2 on the value scale) anchor the white railing and the lighter-valued colors of the shingles as they climb the walls.

*Jill Pilaroscia is a color consultant in San Francisco, Calif. Photos by Charles Miller. Drawings by Gregory Saldaña except where noted.*

# Decorative Sidewall Shingles

Custom-cut wood shingles make a variety of complex patterns

by Dave Skilton

In the summer of 1979, I was struggling to finish a house that we already occupied. The time had finally come to think about the exterior, which we had sided hastily two autumns before using ⅜-in. textured plywood as both a shear panel and a temporary finish. We had a look in mind—the silver-grey of redwood shingles weathered by the salt air—which was the signature of Manchester, our northern California coastal town. The more I looked at the local houses, the more I noticed fancy shingle work, and I liked its subtle beauty. It didn't take me long to decide to use a decorative pattern in the gable ends of our house. That was to be my incentive—a reward for getting the rest of the walls done. Since then, I've completed a number of decorative shingle projects.

**Sheathing**—Decorative sidewall work, like roof shingling, can be installed over "skip sheathing" (horizontal 1x4 boards spaced a

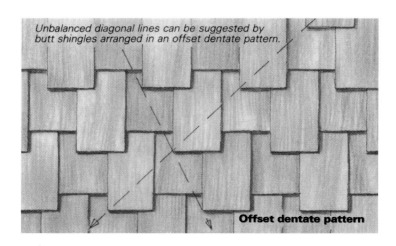

Unbalanced diagonal lines can be suggested by butt shingles arranged in an offset dentate pattern.

**Offset dentate pattern**

board's width apart) or solid sheathing, such as planks, plywood or waferboard that is at least ½ in. thick. No matter what type of substrate you use, it should have a breathable water-resistant membrane under it to allow air circulation behind the shingles. Asphalt saturated 15-lb. felt is a commonly used membrane material, as are the new air-infiltration barriers, such as Tyvek and Typar. Because

local requirements vary, you should check with your building department to find out what membrane materials have local approval. On plywood, the membrane goes on the outside. With skip-sheathing, you need to place the membrane between the sheathing and the studs.

The first course of shingles is doubled because the gap between shingles must always occur over the body of another shingle. The rule of thumb is that the gap should occur no closer to the edge of the underlying shingle than one-fifth the width of that shingle. Nails (4d or 5d hot-dipped galvanized box) should be placed about ½ in. from the edges of the shingles. No nails should be visible on a finished shingle job.

I don't have room in this article to delve deeply into the basics of sidewall shingling, but if you want to know more about the subject, get a copy of the *Design and Application Manual for Exterior and Interior Walls*. It's free

Drawings: Christopher Clapp

**Typical shingle shapes**

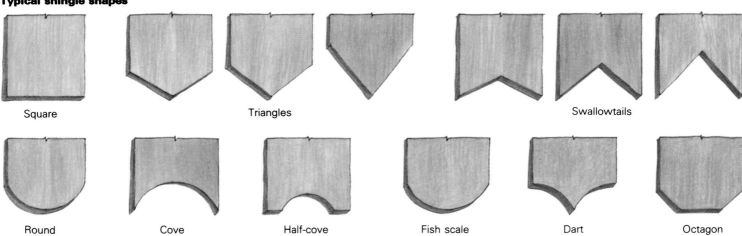

Square

Triangles

Swallowtails

Round

Cove

Half-cove

Fish scale

Dart

Octagon

**Shingle patterns**

*Complex textures can be made by offsetting shingles of a single shape (right column below), such as lozenges, hexagons and the tessellating pattern called propellers. By adding a shingle with a second or third shape (left column below), the possibilities are broadened.*

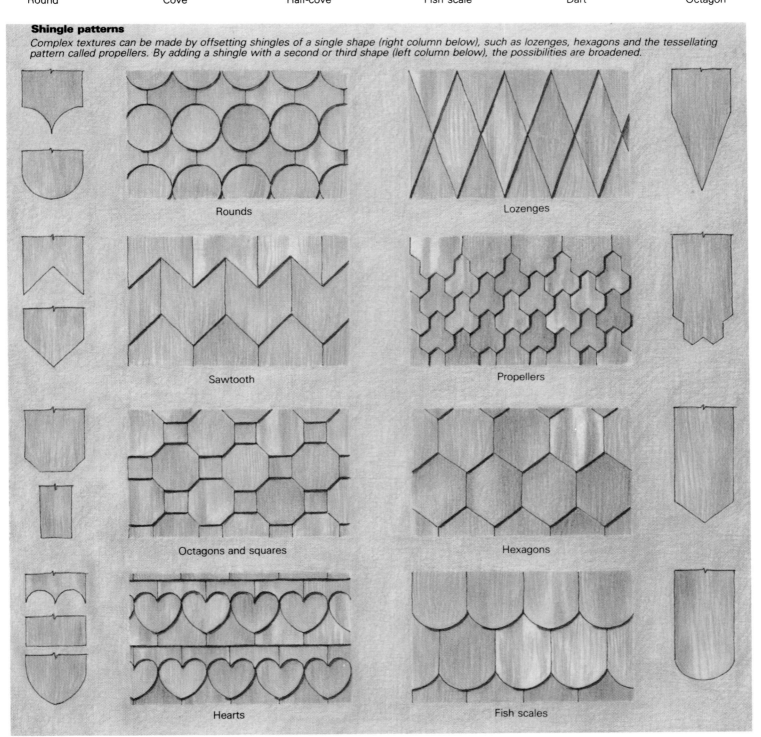

Rounds

Lozenges

Sawtooth

Propellers

Octagons and squares

Hexagons

Hearts

Fish scales

to designers and builders from the Red Cedar Shake and Shingle Bureau (515 100 16th Ave. N. E., Suite 275, Belleview, Wash. 98004). You can also refer to *FHB #28*, pp. 63-67.

**The three variables**—Decorative shingling partakes of all the problems of any sidewall work, plus some special ones. In designing a decorative pattern there are three main variables to manipulate: exposure, width and shape of the shingles. It sounds boringly simple. But subjected to the folk imagination of ingenious carpenters over the centuries, these three variables have produced an astonishing number of patterns.

Exposure refers to the vertical amount of shingle that is "to the weather." This naturally varies with the length of the shingles and the designer's intent. On standard 18-in. shingles the exposure is normally 5 in. to 8 in., but it can be as little as ½ in. or less in the case of doubled or tripled coursing. At the other extreme I have seen so-called "barn" shingles that were 3 ft. long. They allow very generous exposures. Some patterns vary the exposure from shingle to shingle in a given course, making a dentate line. Others vary the exposure randomly for a kind of crazy-quilt effect. In theory the exposure approaches its upper limit at half the length of the shingle, but in practice it is generally held closer to a third.

Width refers to the horizontal measurement of the installed shingle. In many decorative patterns this is held constant because all the laps can then fall precisely over the middles of the shingles in the previous course. But just as with exposure, the width of the shingles can easily be varied within a pattern. The most important thing to keep in mind here is whether the courses lap properly to provide a waterproof wall.

Shape is, of course, what most often comes to mind at the mention of decorative shingling (drawings previous page). Many of these shapes are available precut, and you can even get them in preassembled panels (see sidebar right). Beyond buying precut old favorites, which offer a vast range of combinations, you can cut decorative ends on standard wood shingles and devise one-of-a-kind patterns. That's what I prefer to do.

If you decide to make your own shingles, there are a couple of things to keep in mind. First, if the bottom of a pattern is contained on one shingle, the top will be split between two shingles (bottom drawing previous page). For some patterns the shingles can be identical, and just the act of offsetting them creates a repetition of the pattern. This is true for lozenge and hexagon patterns. Other patterns, such as hearts, and octagons-and-squares require two different shingles. Second, keep in mind that complicated shingles with isolated tongues of wood that are crossed by the grain are highly susceptible to breaking off. Even if you succeed in installing them whole, the stresses of sun, rain, wind and frost will eventually do them in.

**Roof pitch and shingle pattern**—One of the primary design considerations that I discovered on my first gable project was the relationship of exposure to width. I chose a tessellating pattern of hexagons because I was keeping bees on my place and I was intrigued with the structure of honeycomb. However, the pattern proved poorly suited to the low, 4:12 pitch at the gable end. Hexagons establish a strong set of diagonals at 30° to the horizontal, and my roof's slope was 18°. Although I made sure my first course fit in whole-shingle increments across the width of the wall, subsequent courses didn't, and the resulting fragments made the finished gable look a little awkward. Hexagonally-based patterns work far better on gables of about 7:12 pitch because the resulting angle of the roof is more in harmony with the 30° angles established by the hexagons. Matching the apparent diagonals created by your pattern to the roof pitch (or for that matter intentionally opposing the diagonals), can add to the repose or drama of the finished project.

You don't necessarily need shingles with pointed ends to create the effect of diagonals in a wall. By simply offsetting square-butt shingles in a dentate pattern, you will impart a subliminal diagonal line to the wall (drawing, p. 24).

**Cutting decorative patterns**—When I manufactured my hexagonal shingles, I used a wooden jig on a table saw (left photo, facing page). After ripping all my shingles to a common width, I held the shingles against the diagonal stops of the jig and made second and third passes on the saw to add the point to each shingle (for more on sliding jigs for the table saw, see *FHB #53*, pp. 58-61).

I think the best tool for cutting curves in shingles is a bandsaw with a narrow blade,

although I've used a jig saw with a sharp, fine-toothed blade when I had to. To cut the dart-shaped shingles used in the heart pattern (right photo, facing page), I used a 2½-in. hole saw mounted in a drill press. While it's tempting to try and stack the shingles so that a bunch of them can be cut at once, it doesn't work well. Because they are wedges, each shingle in the stack ends up a little different than its neighbor.

Many tessellating patterns, such as the one I call propellers (bottom drawing, p. 25), are based on hexagons and require only one kind of shingle. This greatly simplifies installation and can still create a highly dramatic effect. A single band of special work in an otherwise static pattern can be very effective for emphasizing transition lines, such as the head-levels of windows and doors. Also, if you are creating a pattern that is not repetitive, consider if it might need centering or other special placement for best effect.

The possibility also exists of larger patterns where, for example, a wavy line might run the entire length of a wall in a given course of shingles. In this case, lay out the shingles on the floor as a unit, and then mark them for cutting. Avoid the temptation to move individual shingles up or down too far in place within a given row, unless you are prepared to live with the gaps under shingles that this will cause. Because wood shingles are wedges, they become out of plane with one another as soon as you move them up or down. Some elaborate patterns intentionally exploit this fact, so that the pattern bulges away from the wall.

**Installation**—A straight length of 1x, tacked up level as a guide, is a valuable aid in putting up courses of decorative shingles. I don't recommend snapping a chalkline; it's easy to fall out of register with a chalkline, and the chalk can stain the porous woods used for shingles.

After your hammer or nailer, an accurate square is the most important tool to have along on a fancy shingle job. The problem that seems to come up most often for me with decorative shingles is that they sneak out of plumb rather easily. Sometimes they have only a single point that rests against the guide. An out-of-plumb shingle can cause a series of compounding problems further down the line, along with much cursing and teeth-grinding. Use the square often to check the plumbness of your shingles.

At the top of the wall there is typically a strip of 1x material (called packing) to which the top courses of shingles will butt. On a gable, trim the tops of the shingles near the ends of each course to fit the angle of the roof and keep the nails as high as you can. When all the shingles are on, a rake board, (plus any other moldings you may want to add), can be applied to cover the joint between the top row of shingles and the 1x packing.

Decorative shingles work in concert with adjacent rows of shingles to build patterns,

***Cutting and fitting.*** **A sliding-table jig for a table saw can make short work of cutting lots of shingles with diagonal points, such as the one being cut in the photo at left. Runners beneath the jig correspond to the grooves in the saw's table. The shingles are held against each tacked-on stop to register the cutting angle. Red cedar hearts emerge on the wall as the author places a row of shingles with half-circle cutouts over a row of pointed shingles (photo above). These decorative shingles are 5 in. wide, and they are centered over a row of square-cut shingles that are also 5 in. wide. Just above the shingles is a piece of 1x packing. It will serve as backing for the rake board that trims the top of the wall.**

but the starter row in a decorative scheme usually has straight butts that blend with the other shingles in the field. The shingles in the starter row, however, are ripped to the same width as those in the decorative rows. This allows the centers of the shaped shingles to fall directly over the gaps between the shingles in the starter row (photo above right). I used 5-in. wide shingles for the gable shown here because it's an efficient width to rip out of a bundle of standard shingles without much waste.

Applying shingles to curved surfaces can be a lot easier than trying to do the same job with regular siding. They are especially well adapted to a horizontally concave surface, such as a scooped mansard, because bowing the shingle inward brings it down even tighter on the preceding course. By the same token, a horizontally convex surface is not a good candidate for shingles, because the unnailed end will tend to ride up. Surfaces curved in or out vertically can take shingles equally well, the limit being in the tightness of the arc: the more a shingle must bend parallel to its grain, the more likely it is to split. The flexibility of shingles can be helped along by soaking them in cold water overnight, or by using those with a flat grain. Blunting nails so that they crush rather than split their way through the wood is also help-

ful, particularly when you have to drive a nail close to the end of a shingle.

**Finishes—**In Victorian times paint was the order of the day for everything from sofas to shingles. Unfinished shingles and furniture signaled rusticity, a condition to be tolerated only in cabins and lodges. Nowadays weathered, stained, or clear-finish woods are popular even in urban neighborhoods. Just think of all the pieces of cheaply made turn-of-the-century furniture that people have laboriously stripped and put in their living rooms as precious antiques.

Today, shingles are usually made from cedar or redwood, which are tough woods that will last a long time even without treatment. It is true, however, that a finish will extend the life of shingles, especially those exposed to harsh sun. There are finishes that will preserve the natural or weathered color of the wood, and there are also ways to speed up the weathering of shingles. Redwood painted with a vinegar and water solution, for example, will blacken quickly in sunlight. Or you can simply let shingles weather naturally to the desired color before finishing them clear. The natural preservatives in the shingles should be allowed to leach out for a year before you apply a finish.

While I like the look of weathered shin-

gles, I learned a lesson about trim color while working on my own house. I chose a trim paint that complimented the raw color of the redwood shingles. But as the unpainted shingles weathered, the color combination got so ugly that I had to repaint all the trim.

Something I would like to try someday is staining individual shingles before putting them up, thus adding color to the variables in a pattern. To some extent this is already possible because of the variability of the natural color in shingle woods. But this brings up a potential problem to keep in mind: an inadvertent patch of chocolate shingles in a sea of cinnamon ones can really look like a broken tooth.

There are lots of reasons for choosing shingles, from pragmatic, to aesthetic to historical. For me, the human scale is what appeals the most. One person can put shingles on a house; one person can create a fancy design, from whimsical to elegant. Each shingle is about the size of an open book, and just as readable. Even with regular random-width shingling, each piece requires attention and judgment, fostering a satisfying intimacy between the builder and the project. □

*Dave Skilton is a former builder who is currently studying architecture at the University of Oregon in Eugene.*

# Installing Board-and-Batten Siding

## Plan for the windows and doors first

### By John Birchard

**W**hen I built my own house a dozen years ago, I chose board-and-batten siding for several reasons. It's a commonly used siding in my neck of the woods, where rough-sawn redwood planks have long been used to clad barns and houses. The rhythmic vertical lines of board-and-batten siding complement the upright stance of my two-story house. And because I wanted a rustic look, I used relatively inexpensive, rough-sawn 1x12 boards with tight knots and left them unfinished to weather naturally.

In my innocence, I also thought board-and-batten siding would be pretty easy to install. In some ways it is, but there are many pitfalls on the way to a well-done job. My first project taught me how important it is to consider details like flashing and trim before putting the wood on the wall. Subsequent projects have given me a chance to learn how to plan and execute the work properly while at the same time reaffirming my appreciation for the beauty and durability of board-and-batten siding.

**Patterns, materials and preparation**—The name board and batten encompasses several

*Photo: John Birchard*

**Left to weather.** Unfinished and rough-sawn redwood boards cover the author's house.

variations on the basic theme of affixing square-edged boards vertically to a wall and then covering the gaps between them with another layer of boards. The most common arrangement is to apply boards first, followed by narrow battens. The common variants are boards over boards and reverse board and batten (drawing below). I have also seen board-and-batten siding used with very narrow strips of wood to side curved walls.

The quality of the material you use is more important in determining the final look of the building than the material's weather-tightness. Siding, building paper, sheathing and flashings all work together to keep out the elements, so a knot or a crack in a board isn't enough to compromise the entire assembly. Knotty, rough-sawn wood will work fine for a stained or unfinished exterior as long as the knots are sound. If you want a refined, painted look, you'll need surfaced, knot-free material.

If you plan to paint it anyway, consider using medium-density-overlay (MDO) plywood in place of the boards. You can cover the nailing patterns with the battens, and because MDO has

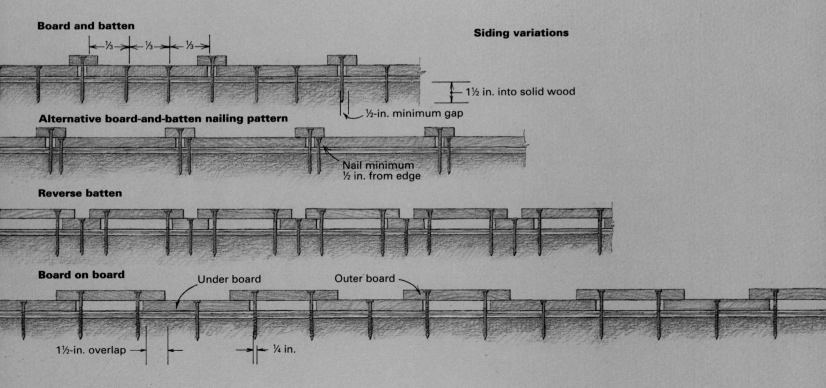

**Board and batten**

⅓ ⅓ ⅓

1½ in. into solid wood

½-in. minimum gap

**Alternative board-and-batten nailing pattern**

Nail minimum ½ in. from edge

**Reverse batten**

**Board on board**

Under board

Outer board

1½-in. overlap

¼ in.

Drawings: Christopher Clapp

a smooth, resin-impregnated skin, it doesn't telegraph the typical plywood-grain pattern when painted. If you're building in an area that is prone to earthquakes, this kind of painted plywood wall can be made to satisfy shear-wall requirements while still looking like board-and-batten siding.

In my experience, redwood and red cedar are the two most common wood species used for board-and-batten siding. In other parts of the country, poplar, Southern yellow pine and cypress are used. Boards are typically 6 in. to 12 in. wide. The battens can be as narrow as 2 in.

Splits and checks caused by the wood expanding and contracting due to seasonal humidity fluctuations are the curse of board-and-batten siding. The best way to ensure the least number of cracks is to use narrow, dry, vertical-grain material, but it's expensive. Flat-sawn boards will expand and contract twice as much as vertical-grain boards. Using affordable wide, green, flat-sawn boards may cause a few more cracks, but they can be minimized with proper nailing techniques (more on nailing later).

I've used green 1x12 construction heart redwood for all of my board-and-batten projects. To compensate for the greenness of the lumber, I order my siding material early in the project. Then I go through it and set aside the best boards for the prominent sides of the building. Finally, I sticker the boards so that they can air-dry as much as possible before they are applied. I cover the top of the pile to shed rain, and I leave the sides open for air circulation. Boards with splits and large knots are ripped into battens and trim.

**Providing a nail base**—Boards should be nailed vertically every 24 in. o. c. In a normal 8-ft. wall, this means that three runs of horizontal blocks are needed between plates. The siding nails must penetrate at least 1½ in. into solid wood. I put the middle row of blocks on edge to double as drywall backing. The other two rows can be flat, making a bigger target and leaving more room for insulation.

Blocking is a job for two carpenters—one to cut the blocks while the other calls out dimensions and installs them. Seeing no reason to lift more than I have to, I prefer to install the blocks after the walls have been raised. I use a pair of 2x4 crutches cut to length to hold the blocks at the right level as they are nailed up.

Another way to anchor board-and-batten siding is to affix horizontal nailers to the outside of the walls. This is the typical fastening method for vertical siding over masonry walls and for post-and-beam buildings. Although I haven't tried it yet, I suppose you could apply horizontal nailers to the outside of a stud wall, thereby eliminating the tedious fitting of hundreds of short blocks. Another plus to this method would be the air space between the back of the boards and the

*Putting up the boards.* Varying the gaps between the boards allows them to be placed in a way that avoids notching the battens or the trim pieces. Full boards flank the doorway, allowing a symmetrical pattern at the entry.

From *Fine Homebuilding* (June 1992) 75:52-56

*Exposed flashing.* A copper head flashing extends beyond the door jambs and tucks into kerfs in the neighboring boards. A 2x head casing with a beveled top edge will tuck under the flashing. The aluminum window has self-flashing flanges. It will simply be picture-framed with 2x trim. The 1x8 siding used here was screwed to ½-in. plywood, eliminating the blocking.

sheathing, which would equalize the moisture content of the boards on both sides and cut down on the tendency for the boards to cup.

You can also use plywood sheathing (½ in. or thicker) as a substrate for the siding (and thereby skip the blocking) if you attach the boards and the battens with screws. Screws should extend at least ¼ in. beyond the inside face of the sheathing, and they should be placed in the same pattern shown in the nailing schedule (drawing, p. 28). Screws should be set with their heads below the surface, so use a pilot-hole cutter to avoid splitting the material (this device cuts a pilot hole for the screw and a countersink for the head in one operation). Incidentally, galvanized screws can bleed black stains into the wood grain because a little bit of plating is abraded from the screw's head by the bit as the screw is driven, leaving bare steel open to corrosion. You can

avoid this problem by spending the extra money for stainless-steel screws (expect to pay about 2½ times more for stainless steel).

**Windows, doors and flashing—**I tack the door jambs and the windows in place before I put up my housewrap or building paper. I wedge the units into the rough openings until they are plumb and square. As I position the windows in the rough openings, I hold a scrap of my siding material against the wall sheathing as a thickness gauge; the outside edges of the jambs should be flush with this piece. Then I drive a couple of nails as pivots near the top of each side jamb (later it may be necessary to swing the bottoms out to get boards behind the ears of the sills).

Next, I tack my tar-paper splines around the bottom and the sides of the windows. The splines should have perfectly straight edges where they

abut the jambs, and they should press tightly against them with their edges curving outward. The blind stops, which are the same thickness as the siding boards, should fit as tightly as possible against the side jambs and be nailed to the framing (top photo, right). Held by the stop against the jamb, the tar paper creates a tight seal without the use of caulks.

Doors and windows that are exposed to the weather need metal top flashing over the heads of their jambs. I prefer a hidden flashing that laps ½ in. over the edge of the window head jamb, then bends inward to the sheathing and up the wall about 3 in. The flashing extends to the outer edges of the blind stops (about 1½ in. past the jamb on either side) so that its ends are away from the space between the rough opening and the jamb. The tar-paper course above the window should lap over the upper leg of the flashing. Then a blind stop is nailed over it with a couple of nails held at least ¾ in. above the fold in the flashing. The head casing hides the blind stop (bottom photo, right). Another way to flash head jambs is with an exposed flashing that runs over the top of the trim.

To install board-and-batten siding around self-flashing aluminum windows, run the building paper over the window flanges, seal any laps with asphalt caulk and notch the boards as needed to fit around the window (photo left).

Flashing the intersection of a gable roof and a sidewall presents a problem with any kind of siding. With board-and-batten siding, the problem is compounded because of the thickness of the battens—you don't want runoff hitting them from the side. My favorite solution is to bend tabs on the bottom step flashing so that any water running down the flashing will be directed away from the wall when it hits the vertical tab (top drawing, facing page). The tabs should be soldered where they overlap, and the horizontal tab should be tapered. With this method, you've got to stretch or compress the layout of your boards so that the vertical tab is placed next to the edge of a board, right next to a batten. If necessary, cut a little notch in the batten to fit it.

**Layout—**The intersection of the battens with the casings has a strong visual impact on the outcome of the project. The most unsightly problem is having to notch a batten around a casing or a sill. Crowding battens at corners or too close to window trim is equally unattractive. It's best when battens land on the upper and lower window trim so that the spacing at each end is equal, but often this perfect balance is unobtainable.

Before I start siding, I use a pencil and a short piece of siding board as a gauge block to mark potential layouts of each board on the wall. By adjusting the gaps between boards, ripping a corner board to a different width or by starting at one end of the wall rather than at the other, I can generally avoid problems. The eye easily overlooks slight variations in the widths of the boards between the battens. On a recent job using 1x12 boards and 3-in. battens, I started with ½-in. gaps between boards. By varying the gaps from ½ in. to 1½-in., I was able to stretch or compress the layout enough to avoid trouble.

***Flashing and blind stops.*** Once the window and the tar-paper splines are in place, blind stops the same thickness as the boards are nailed around the sides and the bottom of the window. The head flashing extends to the edges of the side blind stops (above) and is covered with the top layer of tar paper, a blind stop and the head casing (below).

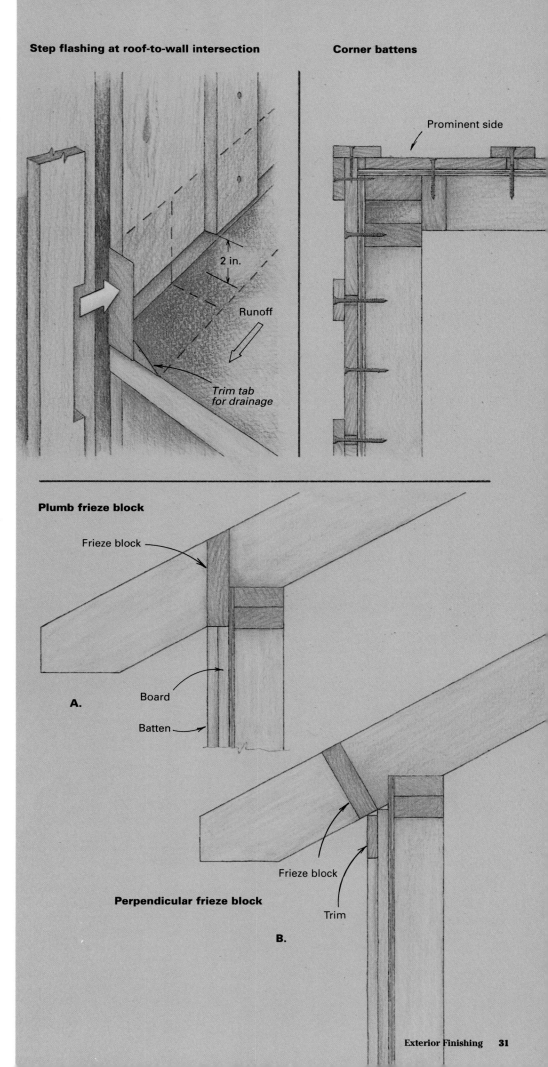

**Step flashing at roof-to-wall intersection**

2 in.

Runoff

Trim tab for drainage

**Corner battens**

Prominent side

**Plumb frieze block**

Frieze block

Board

Batten

**A.**

**Perpendicular frieze block**

Frieze block

Trim

**B.**

At outside corners, the last boards on the most prominent sides of a building should overlap the ends of the boards on the adjoining wall (drawing top right). This puts the joint between the corner battens on the least prominent wall.

**At the top**—Where the walls meet the roof, several different treatments are possible. For an open eave, with the rafter ends showing, I like to set the rafter frieze blocks plumb so that the boards and the battens can butt into them (drawing right, detail *A*). The top edges of the blocks are beveled at the roof pitch while their bottoms are square. This blocking method eliminates the need for notching the boards and the battens to fit around the rafters. Of course, it is also possible to set your blocking plumb in plane with the wall, which allows the boards and the battens to extend to the underside of the roof decking. This requires a lot of careful notching, but the results are very clean if the job is done well.

If you set your blocks perpendicular to the rafters, a trim piece applied below the rafters and over the boards can also be used with an open eave to avoid notching (drawing previous page, detail *B*). For a soffited eave, the boards and the battens both can simply butt up to the bottom of the soffit. But more commonly a horizontal trim piece is applied to the wall along the bottom of the soffit where the battens end.

**Nailing details—**The California Redwood Association (405 Enfrente Dr., Suite 200, Novato, Calif. 94949; 415-382-0662) recommends nailing the boards with one or two nails, depending on the width of the board. For boards less than 8-in. wide, one nail in the center will suffice. Wider boards should be secured with two nails that are equidistant from the edges and each other, dividing the board into thirds (drawing, p. 28).

In an attempt to minimize the number of nail heads visible on the siding, some carpenters nail the boards near the edges where they will be covered by the battens. You can do this if you nail *only one edge*, as shown in the alternative board-and-batten nailing pattern. If you nail the boards at both edges, you will restrain them improperly and run the risk of cracking them down the middle (photo top right).

In the drawing on p. 28, the nailing pattern for board on board and reverse board and batten doesn't divide the board into thirds. Instead, the outer boards lap the under boards by 1½ in., and the nails are located about ¼ in. inside the lap. Because the nails securing the outer boards aren't completely embedded, they will flex a bit to accommodate seasonal variations in the dimension of the boards without splitting them.

The battens are secured with nails at the centerline. This holds the edges of the boards flat against the wall while still allowing them to move. The batten should overlap the board by at least ½ in. If the gap between boards is wide —say 1½ in. or so—I predrill the nail holes to avoid splitting the batten. Another way to avoid splitting battens in this situation is to place a nailing strip the same thickness as the boards down the center of the gap to serve as backing.

It's important to nail the battens on as soon as possible after the boards are applied—especially with green lumber—so that the boards won't cup as they dry. If they are allowed to cup (which can happen fast on south-facing walls), you won't be able to flatten the boards out without cracking them.

I've used hot-dipped galvanized nails on several jobs, and for rustic work I don't mind the small amount of staining that occurs eventually around the nail heads. But lately I've been won over by stainless-steel, ring-shanked siding nails because they grip tenaciously, and their slightly tapered heads allow them to be countersunk without crushing the surrounding wood. They are also the only nails I know of that don't corrode or discolor wood siding.

Before I affix the boards to the wall, I snap chalklines to show the center of my nailing base —be it blocking, rim joists or plates. It's best to

***Wrong.* Installed green with nails at the edges, this cedar 1x12 cracked down the middle as it dried out.**

Photo: John Birchard

***Hard-hat country.* Working in tandem, one person on a ladder and another on the ground can install long boards easily with the help of a lever to hold the boards steady.**

apply siding before doing the plumbing or electrical work to avoid nailing into a pipe or a cable. If plumbing or wiring is already in place, be sure to put nailing shields over the danger zones.

**Putting them up—**Flat-sawn boards should be oriented with the bark side out. This will decrease the incidence of "feathering"—the separating of spring and summer wood that can occur on flat-sawn boards. I know this is contrary to the wisdom that says placing a board with its bark side out will exacerbate the board's tendency to cup and then split. It just doesn't seem to work out that way in practice. Regardless of whether the bark faces in or out, a board on a wall wants to cup outward because the backside retains more moisture than the exposed side.

Your best insurance against splitting is correct nailing. To keep the boards from cupping, back-priming them with whatever finish you're using or a water-repellent preservative is also recommended by the lumber associations.

In most cases, two workers, one with a ladder, the other with a hard hat, can apply board-and-batten siding very efficiently with no need to set up and move cumbersome staging. The boards are cut to length, then stood straight up and lifted into nailing position. For long, heavy boards that reach the foundation, I rig a simple lever that the worker on the ground can operate with his foot to hold the board up while the first nail is driven (photo below left). I usually drive the first nail at the highest block—not the top plate. Once the board is hanging, I check it for plumb and then drill pilot holes for the top nails to avoid splitting the board near its end. To ensure holding power, the diameter of the pilot holes should be no larger than three-quarters of the fastener's shank.

Unless the boards are especially brittle, which can happen with kiln-dried material, I don't bother drilling pilot holes anywhere but near the ends. The boards sometimes want to warp away from the building at their ends, so I nail into both the mudsill and the rim joist to counteract this tendency.

If the boards aren't long enough, I make scarf joints by cutting the abutting ends of the boards at 45°. The upper board laps over the lower one, creating a beveled edge that conducts water away from the joint. Scarf joints should occur over blocking, and the nails should be angled into the blocks through pilot holes. Scarf the battens similarly and stagger the joints so that they don't line up next to each other.

By the way, the Forest Products Laboratory recommends treating the end grain of scarf joints with a water-repellent preservative before the boards are installed on the wall. They also suggest beveling the bottoms of the boards to create a drip edge.

If the windows I'm installing have ears on the sills, I swing them out at the bottom to get the boards in behind. Butt the boards up as tightly as possible to the blind stops, notching them where necessary to accommodate the openings (photo, p. 29). Once all the boards are on the wall, I permanently nail the windows into their rough openings.

The battens and the exterior casings occupy the same layer on top of the boards, so the exterior trim must be installed before the battens that will abut either the apron or the head piece. You can use material of the same width for both casings and battens, or the casings can be made noticeably wider, thicker or both.

The nail heads in battens and boards should be set slightly below the surface for maximum holding power. But don't set them more than ¹⁄₁₆ in. Setting the nails deeper allows water to collect in the recesses. □

*John Birchard is a woodworker in Mendocino, Calif. He is the author of* Make Your Own Handcrafted Doors and Windows *and* Doormaking Patterns and Ideas *(Sterling Publishers). Photos by Charles Miller except where noted.*

# Siding with Clapboards

## Up one side or down the other, apply them correctly

by Bob Syvanen

**M**y dictionary says that the word clapboard probably derives from the Middle Dutch *clapholt*—"clappen" meaning to crack or split and "holt" meaning wood. And in fact early clapboards were made by splitting or riving them from logs, rather than by sawing them. Still another explanation is offered by William B. Weeden in his book *Economic and Social History of New England 1620-1789* (Corner House Publishers, Williamstown, Mass. 01267, 1978): "Bricks were laid against the inner partition or wooden wall, and covered with clay. Boards were placed on the outside, first called 'clayboards,' then corrupted into 'clapboards.'"

Today the word clapboard is used generally to refer to many different kinds of horizontal lap sidings. But here in New England, the word clapboard nearly always means bevel siding (photo below).

**Tools—**The nailing for clapboard work is light, so a 16-oz. wood-handled hammer is right for me. I use a 10-point, 20-in. crosscut handsaw for most of the cuts. And I use a utility knife rather than a pencil for marking. A knife mark is an exact line to cut to, whereas a pencil line varies in width and leaves you guessing as to which side to cut on.

I use a tape measure, story pole and chalk line for setting course lines. To locate nailing, I plumb down the stud locations onto the face of the building paper with a 4-ft. level. A combination square, block plane and a homemade siding gauge round out the list of essential tools for siding work. I use the siding gauge to mark clapboards for a good fit against the trim. When used properly, it keeps adjustments with the block plane to a minimum (more on using the siding gauge in a moment).

**Figuring material—**Most wood clapboards today are cedar, redwood or pine. Common sizes are ½-in. thick (at the bottom) by 6-in. wide, ⅝-in. by 8-in. and ¾-in. by 10-in. All sorts of grades are available, but lately I've been using vertical-grain clear cedar. One surface is smooth for painting and the other is rough for staining. It comes in lengths from 3 ft. to 20 ft., wrapped up in bundles of ten boards each. These days, the price in my area for ½-in. by 6-in. cedar is $.82 per lineal ft.

Some of the best clapboards I know of are made at a small family-run mill in Sedgwick, Me. (see *FHB* #31 pp. 35-37). Their #1 Premium Clears are radially sawn from eastern white pine. Their standard size is ½-in. by 6-in. by 8-ft. maximum length.

The rule of thumb for estimating siding material is to calculate the square footage of the wall area (minus window and door open-

**Beveled clapboards, the quintessential siding, have been a tradition in New England for 300 years. Despite the advent of power saws, such as the radial-arm saw standing idle in the photo below, most carpenters today still cut clapboards the same way their forefathers did—by hand.**

ings) and add 25%. But there is a more accurate way. Multiply the wall area (minus window and door openings) by the area factor shown below for the size of the siding that you are using:

| Siding | Area Factor |
|---|---|
| (assuming a 1-in. lap) | |
| 1x4 | 1.69 |
| 1x6 | 1.40 |
| 1x8 | 1.30 |
| 1x10 | 1.24 |

Whichever method you use, add an extra 5 percent for waste.

I think the best nails for clapboards, if you are going to leave them exposed, are ring-shank aluminum sinkers because they won't ever rust and stain the siding. They're fairly expensive and not a stock item at my lumberyard. Stainless steel nails are also available. But I've used hot dipped galvanized box nails for years with no problems. Trade associations and manufacturers recommend using at least 7d nails for ½-in. by 6-in. clapboards, but as long as I'm nailing over plywood sheathing, I use 5d nails. After penetrating the plywood, 5ds still sink ½ in. into the studs. They have smaller heads than 7d nails and are less expensive. I figure 5 lb. of 5d box nails per 1,000 square feet of 6-in. siding.

**Preparation**—To prevent the wood from cupping, I prime the back sides of all trim and siding, including clapboards. Boards will cup when only the face is painted because moisture will be absorbed through the unpainted backside. The wider the board the more evident the problem. I've seen 12-in. corner boards on a three-year-old condo that looked like barrel staves because they were only face-painted.

Corner boards, window casings and door casings are often wider than the studs in the wall behind, leaving no nailing for the clapboards that butt into them. To provide nailing, I fill in around corners and rough openings at the framing stage with 2x4 or 2x6 studs laid flat. I run the corner boards wild at the bottom and trim them later to fit the bottom of the clapboards.

You have to take precautions to keep water from running behind trim. Inside and outside corners and openings for doors and windows are the main areas of concern. Before nailing on trim, I staple up strips of 15-lb. builder's felt, which I call splines. I make them long enough to extend 3 in. above headers and 6 in. or 8 in. below sills. At least 4 in. of felt should show at the sides of all trim (drawing, p. 36). I cover the rest of the wall with building paper as well and tuck it behind these splines except at headers, where it goes *over* the splines.

I flash the head casings with copper, zinc or aluminum. This flashing should extend over the felt splines at the top corners of the windows and doors. I use rolls of 8-in. wide zinc most of the time because I like the way it works, but I use copper for really first-class work. I cut a strip as long as the head casing,

The course lines for clapboards are marked on a story pole. The pole can then be held against the frieze board anywhere around the house in order to transfer the lines to casings and outside or inside corner boards. *Photo by Pat Syvanen.*

then I cut it lengthwise with a utility knife, giving me a 4-in. wide strip. I bend the flashing over a 2x4 and space the bend so that ⅛-in. of flashing will extend beyond the casing. After nailing it in place and applying the clapboards, I use a 2x4 block and bend the extra ⅛-in. of flashing over the edge of the casing.

At the bottom corners of windows and doors, the 6 in. or so of exposed spline should run behind the first course of siding below the sill and over the top of the second course below that. I trim the excess with a utility knife.

On the bottom of the window sills, I cut a ½-in. deep rabbet wide enough to receive the upper edge of the clapboard. Most windows come with this rabbet, but I check to make

sure it fits the clapboards I'm using. If you have to cut or widen it, you're better off doing it before the window is installed. A circular saw with a rip guide will do the job nicely. It can be done after installation, using a utility knife and a chisel, but it's not easy.

To protect the space under the sill, I cut a piece of felt 8-in. wide and about 6-in. longer than the sill. I tuck this behind the side splines and hard up into the rabbet. The bottom goes over the paper below.

At the top of the wall, I rabbet the bottom of the frieze board so that the top edge of the top course of clapboards fits behind it. I use a router for this and make the rabbet ½-in. deep by whatever width it takes to receive the upper edge of the clapboard. I pack out rake

After the builder's felt is applied, a starter strip, representing the amount of overlap between courses, has to be nailed up so that the first course of clapboards will lay at the same angle as all the rest. Use 2-in. rippings from the top edge of the clapboards.

boards with 1x3 furring, or thicker stock if necessary, depending on the thickness of the clapboards. I nail them flush with the top edge of the roof sheathing (on houses without overhangs) and then nail the rake boards into them. This leaves enough space to slip the clapboards neatly under the rake.

**Layout—**I try to lay out courses so that the bottom edges line up evenly with window sills, header casings and frieze boards. This looks good and minimizes the number of boards I have to notch. I vary the exposures slightly if I have to—¼-in. more or less isn't noticeable, especially when spaced out over several courses. When I can't work it out exactly right, I choose the least objectionable option. If the height of door tops is different from window tops, I lay out the siding even with the window tops because there are more of them. When window tops are a few inches below the frieze board, I lay out the siding to avoid a narrow course under the frieze. An even better solution here is to use a wider frieze board that eliminates the need for a narrow strip of clapboard over the windows.

The maximum exposure for a 6-in. clapboard (it actually measures 5½ in.) is 4½ in. The exposure cannot be increased because the minimum 1-in. overlap at the upper edge would be lost. If I have a window opening that's 56 in. high, I divide 56 in. by 4½ in. The result is 12.44, which I round off to 13 and divide into 56 again. I come up with an

exposure of slightly more than 4¼ in. for each course. In general, there is a lot of fudging of course exposure, and the time to do it is when laying out the walls.

To keep layouts consistent around the building, I make up a story pole that reaches from the bottom of the frieze board to a few inches below the top of the foundation. The frieze board is the reference point from which all of the layout marks are made (photo on previous page). In addition to

At inside corners, a square stick, usually ¾-in. by ¾-in., is nailed up and the clapboards butt into it.

clapboard course lines, the story pole shows door casing heights, the bottom of window sills and the top of window casings. The story pole can then be used for layout on the gable walls as well by aligning it with the tops of the window casings.

The course marks on the story pole are first marked to indicate the butts or bottoms of the clapboards. However, I start at the bottom of a wall when applying clapboards, setting them to a chalk line locating the upper edge. So to make the story pole work, I add the width of the clapboard to each butt mark. To avoid confusing these marks with the other layout marks, I make them very dark. Transferring the measurements from the story pole to corner boards and door and window casings is the last step of preparation.

**Clapboarding—**After the lower portion of a wall has been covered with felt, a beveled starter strip, 2-in. wide and the thickness of the top edge of the clapboard is nailed in place (photo above). I like to let it lap about ½ in. over the top of the foundation. The first clapboard course is laid to a snapped chalk line that locates the top edge of the clapboard. Traditionally, clapboards were applied from the top course down (some carpenters still do it that way) because early clapboards were hand-split and the top edges were not parallel to the bottom. The trued-up bottom edge became the control edge, and applying from the top down was the only way clapboarding could be done. If

**Preparation and installation details**

4-in. wide metal flashing
folds over front of casing

Corner board

Course marks

Felt spline

*Course marks are
transferred from
story pole onto
splines along corner
boards and casing.*

Casing

Sill

Drip
kerf

Rabbet

Clapboard

*Side spline runs behind first
clapboard below window,
on top of second.*

*15-lb. builder's felt
slipped under splines.*

Chalk lines

Clapboard

Strips of felt behind
joints in clapboards

Starter strip

Drawing: Michael Mandarano

you're using scaffolding that attaches to the wall, then working from the top down eliminates the hassle of leaving out clapboards and filling them in later. To work from the top down, I snap lines corresponding to the bottom edge of the clapboard. The uppermost board goes on first, and I nail it near the top edge. That allows me to slip the second course under the bottom edge of the first, until it's on the chalk line, and nail through both courses. I follow the same procedure all the way down.

To get good fits at corner boards and casings, I hold the clapboard in place, straddle it with my siding gauge, and mark it with the utility knife while the gauge is held hard against the trim (cover photo). I like a snug fit at the trim, but not so tight as to push the trim around. This is important around vinyl windows that have no wood casings. They are very flexible, and the narrow side casings are easily pushed to accommodate a piece of clapboard that's too long. I use a block plane to adjust bad or tight-fitting joints.

When there are many courses of a specific length of clapboard, I cut one piece to fit and then check it at each course above to see where else it will fit. When I'm lucky, I can use the fitted clapboard as a pattern for all. Even if the pattern works for only half the courses, duplicate cutting is a time-saver. It works particularly well with very short lengths of clapboards, where I can set up a stop block and cut them on the radial-arm saw or table saw.

All butt joints between clapboards in the field of a wall should occur over a stud, with each end nailed to that stud. To make the joint leakproof, I put a 2-in. wide strip of building felt (I cut a whole mess of them beforehand) behind this butt joint. The traditional way to make this joint is by shaving or skiving the clapboard ends so they overlap by around 2 in.

I nail into each stud about ½ in. above the butt of the clapboard so that the nail also goes through the top of the course below. This is another area in which I disagree with the manufacturers and trade associations, who recommend nailing just above the course below. The reason they call for this is to allow room for the clapboards to expand and contract. But when I'm using 6-in. clapboards, the exposures are around 3½ in. to 4 in., and with the clapboards painted on both sides, there is no movement resulting from moisture problems.

Nailing above the top edge of the course below can result in splitting and cupping if you hit the last lick too hard, particularly with red cedar. Sometimes the bottom edge of the clapboard will cup away from the course below, which looks bad and doesn't seal the joint. But the main reason I nail through both courses is that, if the clapboards are going to be painted, I always set the nails so they can be filled later. And if you try to set nails in clapboards where they have no support directly behind them, the clapboards will split.

To fit clapboards around a round window, the clapboards are laid out on the ground and a pattern of the window, made of building felt, is traced onto them.

**Corners—**At inside corners I butt the clapboards against a ¾-in. square corner stick (bottom photo, p. 35). The size of the stick depends on the thickness of clapboards and exposure. I nail it over a creased spline of 15-lb. builder's felt.

There are two choices for the treatment of outside corners. I prefer butting the clapboards against corner boards, but you can miter the clapboards as well. To miter them, hold the clapboard in position and mark the backside at the corner. Set the clapboard in a miter box (a site-built miter box tall enough to accommodate the clapboard works best), shim the bottom out equal to the lap over the course below and make a 45° angle cut. Adjust the fit with a block plane. Cross nail through the miter with 4d galvanized finish nails.

**Gables—**I like to tuck the clapboards behind rake boards rather than fit them to the bottom edge of the rake. To determine how thick the packing strip must be, I stack a few pieces of clapboard to simulate the wall, and then measure. For ½-in. by 6-in. clapboards with a 4-in. exposure, a ¾-in. packing strip works well.

If you have to side around a gable vent, arch-top or round window, it's easiest to make a pattern out of building felt and lay out the cuts on the ground (photo above). It saves a lot of last-minute ladder-top fitting.

**Finishing up—**I set nails as I go because if a clapboard should split then, it's easy to replace. If I have to remove a clapboard after courses have been nailed in place above, I drive the nails through with a large nail set. Since the nails in the course above go through the top edge of the board I'm replacing, I drive them through this board as well. A new board is set in place and nailed. The bottom edge of the course above is renailed and the old nail holes filled.

Setting all the nails is extra work, and a lot of people don't bother. But if the clapboards are going to be painted, they look much better if the nails are set and filled. Paint doesn't hold up well over galvanized nail heads. I fill all nail holes with Old Time Putty (Sterling Cark Lurton Corp., Malden, Mass. 02148), taking care not to dimple or dent the filled hole.

I have seen carpenters push putty into nail holes, cut it by sliding the hand to the side and then wipe the residue with a finger. The result is a dent at the hole with putty all around it. When the house is painted, every nail location is accented. To avoid this, I take a wad of putty in one hand, push it firmly into the hole and then slice if off at the surface of the clapboard with a clean putty knife.  □

---

*Consulting editor Bob Syvanen is a builder in Brewster, Mass. Photos by author except where noted.*

# A Quality Deck

## Built-in benches, recessed lights and a spacious spa highlight a high-class deck

by John Baldwin

Some time ago, Chris and Judy Lavin hired our small design/build firm to craft a fancy cherry wet bar and remodel the kitchen in their 1950's home. The completed work enhanced the home's interior, but the exterior was another matter. For one thing, its T&G cypress siding was cupped, checked and shedding paint. For another, the existing doors and metal casement windows had suffered from decades of New England weather. Equally significant, the ample backyard was unmanicured and unused.

Based on the quality of our earlier work, we were commissioned to replace the existing cypress siding with oiled 1x6 V-groove Western red cedar, and the old windows and doors with new wood ones. The heart of the job, though, would be in the backyard. That's where the Lavins wanted us to build a 1,200-sq. ft. hot-tub deck. At least, that's what they called it. When we looked over the drawings supplied by architect John Gardner Coffin, we discovered that the so-called "deck" would be a lot more like a piece of built-in furniture than an exterior add-on (photo right).

**The plan**—Attached to the north side of the house, the deck was to be accessible from the kitchen and living room through a series of sliding-glass doors. Instead of the usual handrail, it would be enclosed by a continuous 21½-in. high bench. This substitution was acceptable to the local building department because the bench would be relatively wide (about 18 in.) and because the surface of the deck would be relatively low to the ground (5-ft. maximum above grade).

The bench would be clad on both sides with vertical 1x6 cedar siding, which would mesh visually with the newly refurbished exterior of the house. Enhancing this effect, the cedar on the outboard side of the bench would extend below the rim of the deck to within about 2 in. of grade, forming a skirt that would conceal the deck framing and concrete piers.

Coffin also called for a set of stairs to fan out from the deck into the newly sodded backyard. A second stair, adjacent and parallel to the house, would allow easy access to the east end of the yard.

After discussing the project at length, the Lavins decided to equip the benches and adjacent house soffits with recessed low-voltage lighting so that the deck could be used at night. They also settled on an 88-in. square Sundance "Cameo" spa (Sundance Spas, 13951 Monte Vista Ave., Chino, Calif. 91710; 714-627-7670) as the deck's centerpiece, which at the time was the biggest four-seater on the market (photo, p. 41). Finally, they requested that a built-in planter be incorporated into the perimeter of the deck, and that several more planters be built on wheels so that they could be moved around.

For longevity, Coffin specified that the deck be supported by poured concrete piers and the bottoms of the two stairs be supported by concrete-block foundations, with the piers and foundations extending below the frostline (42 inches in this part of New England) to prevent frost heave. He also called for the use of pressure-treated Southern yellow pine for the framing and decking (though we amended the decking material later) and the use of galvanized nails and metal connectors throughout.

**Trenches and craters**—Even before we broke ground, we encountered our first snag: as designed, the new deck would cover the existing septic tank. The Connecticut Wetlands Department (whose domain seems to include any land that ever gets wet) required a minimum clearance of 10 ft. between the deck and the tank. This meant that at considerable expense, we would have to bypass the old septic tank and install a new tank and leach field the prescribed distance from the deck. We were also required to excavate a 1,000-gal. drywell for the spa. We informed Chris that "we can build your deck, but your backyard's gotta go." For-

tunately, as a former builder, he understood Murphy's Law and gave us a quick go-ahead. The project started in late June, the beginning of the hottest summer Connecticut had seen in 30 years.

With the septic system rearranged, we set batter boards and ran string grids to plot the centers for the 24 poured-concrete piers. We used powdered lime to lay out the locations of the two buried 8-in. concrete-block walls that would support the stair stringers. The holes and trenches were dug with a backhoe.

After forming the 8-in. by 16-in. concrete footings that would support the block foundations, we shoveled 1½-in. to 2-in. crushed stone into all the trenches and pier holes, forming level pads 42 in. below grade. The tops of the piers would extend 8 in. above grade, so we formed them with Sonotubes—cylindrical cardboard tubes available from masonry suppliers.

Before pouring the concrete, we determined the finished elevations of the piers using a transit level, and then cut each Sonotube about 10-in. short. During the pour, we rested the Sonotubes on the gravel pads, filled them

From *Fine Homebuilding* (August 1991) 69:46-49

Clad with Philippine mahogany and Western red cedar and washed by incandescent light, the deck is more a piece of built-in furniture than a simple platform.

roughly one third full of concrete, and then pulled them straight up 10 inches. This let a good measure of concrete slump out the bottoms of the tubes so the footings and shafts of the piers would be cast in a solid block.

Once the Sonotubes were filled with concrete, we set a ½-in. by 12-in. anchor bolt in the center of each one and troweled the tops smooth. The anchor bolts allowed us to bolt a TECO galvanized post anchor (TECO Products, P. O. Box 203, Colliers, W. Va. 26035; 800-438-8326) to each pier. These anchors have a cam-lock design that allows more than 1 in. of play in positioning the deck posts, while at the same time raising the posts about 1¼ in. above the tops of the piers to prevent the posts from wicking moisture from the concrete. Once we laid up the block walls and backfilled around the piers and walls, we were ready to start framing.

**Basic framing**—The pressure-treated framing for the deck consists of 4x4 posts and built-up girders (doubled and tripled 2x8s) supporting 2x8 joists spaced 16 in. o. c. For

drainage, we sized the posts so that the deck would slope away from the house ¼ inch per 4 feet. At the house, the joists are supported by a 2x8 ledger fastened with galvanized lag bolts and lead shields to the existing concrete-block foundation.

Because its base would be positioned 16 inches below the surface of the deck, the spa needed a platform of its own. We built a 7-ft. by 10-ft. platform that also allowed access via a hinged desktop hatch to a control panel mounted in the side of the spa (more on that later). Supported by concrete piers and 4x4 posts, the platform was framed with 2x8 joists spaced 16 in. o. c., then topped with 1¼-in. by 5½-in. pressure-treated pine. This was sufficient to support the more than 5,000 lb. the spa would weigh when full of water and hot-tubbers. A short knee wall built around the pe-

rimeter of the spa platform supports the edge of the deck above.

The stair stringers consist of 2x10s that rest on 2x6 pressure-treated plates anchor-bolted to their supporting concrete-block foundations. We laid out the stairs so that each riser is the height of a single deck board (5½ in.) and each tread is three deck boards deep. To reinforce the longer stringers adjacent to the house, we nailed a 2x4 to the side of each one.

**The bench chassis**—The bench framing consists of two short 2x4 frame walls—an inner one and an outer one—capped by a third one laid flat (drawing, p. 41). We aligned the outer walls so that the outside faces are plumb with that of a continuous, horizontal 4x4 beam lag-screwed to the outsides of the deck posts. This allowed the outer wall and the beam to serve as a nailing base for the vertical cedar siding. Of course, positioning the outer walls of the bench this way placed them outside the rim joists of the deck instead of over them. To support the outer walls, we nailed a 2x4 ledger to the outboard sides of the rim joists.

Also, where necessary we shimmed the 4x4 beam straight.

Where the bench runs perpendicular to the deck joists, its inner walls are secured with 16d nails to the joists. Where the bench parallels the joists, its inner walls are supported by 2x8 blocking installed 16 in. o. c. between the joists. To ensure that the bench would be as straight as possible, we cut its framing components accurately by clamping a stop to the fence of our miter saw.

We left one section of the bench open temporarily. That allowed seven of us to slide the spa from the flatbed trailer when it was delivered and lower it onto its platform. The spa was then plumbed and wired before the decking was installed.

**Switching to mahogany—**With the extensive deck framing nearly completed, everyone was tired of coping with pressure-treated Southern yellow pine sawdust and splinters, yet we still had to install 1,200 bd. ft. of decking. One of our carpenters had recently built door jambs out of Philippine mahogany, and he suggested that we use mahogany instead of pressure-treated pine for the decking.

Because Chris had been designing, building and racing power boats for years, he knew that the durable, rot-resistant wood was commonly used for boat decks and hulls. He also liked its appearance—the color of Philippine mahogany ranges from deep maroon to almost pure white, and its grain figure from quarter-sawn to bird's-eye. Southington Specialty Wood Company of Southington, Connecticut, offered us the mahogany at an attractive price, and because they were nearby, their shipping costs were reasonable. To everyone's delight, Chris opted for the mahogany, although it added about 10% to the cost of the deck.

Before we started on the decking, we secured a length of ¾-in. plywood to the bed of our 10-in. miter saw. This plywood raised the 1¼-in. by 5½-in. deck boards far enough off the miter-saw table that the sawblade could cut through the stock in one pass. To help support the ends of the decking, we extended the plywood about 2 ft. past either end of the saw, leveling the plywood wings with wood blocks. Finally, we nailed a narrow wood strip on top of the plywood about 5 in. to either side of the sawblade, perpendicular to the fence. These fulcrums raised the ends of the deck boards just high enough to produce a 2° back bevel for a tight fit at the butt joints.

We soaked the end grain with CWF wood preservative (The Flood Co., P. O. Box 399, Hudson, Ohio 44236; 800-321-3444) before installation. Originally, we had planned to screw and peg the decking to the joists, but decided to use 12d spiral-cut galvanized nails instead to cut costs. The mahogany took nails extremely well; whenever a board split it was invariably due to a preexisting check. To avoid damaging the wood with hammer marks ("elephant tracks"), each of us knocked a small knot out of a shim shingle so that we could slip the shingle over the nails before driving

*Lighting the basement.* The basement was originally lit by casement windows housed inside a concrete areaway. To prevent the deck from dimming the basement, removable mahogany grates (photo above) were installed directly over the areaway.

*The access hatch.* The spa's control panel (photo left) is reached by way of an access hatch attached to the deck with a stainless-steel piano hinge.

*Planting the perimeter.* At the client's request, a planter (photo below) was built into the deck's perimeter. It's enclosed by a bench topped with mitered mahogany decking.

## Bench framing

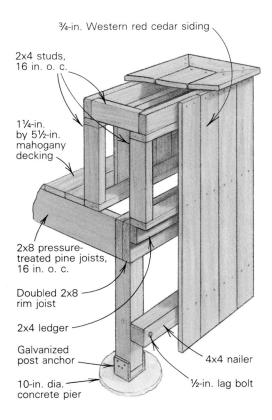

- ¾-in. Western red cedar siding
- 2x4 studs, 16 in. o. c.
- 1¼-in. by 5½-in. mahogany decking
- 2x8 pressure-treated pine joists, 16 in. o. c.
- Doubled 2x8 rim joist
- 2x4 ledger
- Galvanized post anchor
- 10-in. dia. concrete pier
- 4x4 nailer
- ½-in. lag bolt

The Sundance "Cameo" spa weighs more than 5,000 lb. when full of water and hot-tubbers. It's supported by its own platform, 16 in. below the surface of the deck.

them home (after all, nobody's perfect).

The 25-in. by 45-in. access hatch next to the spa (bottom left photo, facing page) consists of 2x6 pressure-treated pine joists spaced 16 in. o. c. and topped with the mahogany decking. The heavy hatch was originally designed to be lifted up out of its opening, but this would have invited back problems. Our solution was to hinge the hatch to the deck with a 45-in. long stainless-steel piano hinge fastened with stainless-steel screws.

**Making the grates**—Before the deck was built, Chris's basement workshop was illuminated by two casement windows housed inside a 2½-ft. wide by 12-ft. long concrete areaway (a sunken space that allows air and light into a basement) adjacent to the house. To prevent the deck from blocking this light, we installed three removable 2½-ft. by 4-ft. mahogany grates directly over the areaway (top photo, facing page).

The grates consist of 1¼-in. wide by 1⁹⁄₁₆-in. thick mahogany strips ripped out of leftover deck boards, lap-jointed into a grid-like pattern and supported by 2x4 ledgers. We cut the lap joints using a table saw equipped with a dado blade and a 40-in. long wood fence C-clamped to the saw's standard miter gauge. A small wood guide pin notched into the bottom of the wood fence was positioned so that after each lap joint was cut, the workpiece was repositioned with the newly cut joint stopped by the pin the correct distance from the next cut. This ensured perfect, uniformly spaced lap joints. We nailed the grates together with 6d galvanized finishing nails, then used a router

to round over all 408 openings in the grates. The finished grates let plenty of light into the basement, and each is sufficiently rigid to support several hundred pounds.

**Mitering the bench tops**—The plans called for the use of three parallel deck boards for each segment of the bench top, with the boards crosscut at each end. For a more elegant look, we crosscut the center boards only, capping the end grain of each one with a short length of mahogany mitered to the two outside boards (bottom right photo, facing page).

To prevent splinters and to soften the bench visually, we rounded over the ends of each center board using a router fitted with a ⅜-in. roundover bit. We laid out the miters in place by lapping the two boards to be mitered, marking their intersection at the inside and outside edges, and then scoring the cut line with a straight edge and a utility knife (where the benches met at a 45° angle, the miters were more than 11 in. long). We cut the miters with a circular saw and fine-tuned the joints with a smoothing plane, giving each miter a shallow back bevel for a tight fit at the top. Before nailing down the bench tops, we installed 2x10 pressure-treated blocking beneath the location of the mitered joints and smoothed the top of the bench framing with a power plane.

**Lighting it up**—Before the benches were boxed in with cedar siding, they were wired to accommodate several lighting fixtures. Once the siding was applied, we cut small rectangular openings in it to house the fixtures.

Coffin called for the use of Prescolite model

37G-1 lighting fixtures (Prescolite, Inc., 1251 Doolittle Dr., San Leandro, Calif. 94577; 415-562-3500). These fixtures each hold a pair of 25-watt incandescent light bulbs concealed behind a plastic louvered grille that sheds weather and directs the light downward toward the mahogany decking. Combined with recessed lighting in the house soffits, the bench lights provide plenty of glare-free illumination for socializing.

**Finishing off**—In addition to a fixed 4-ft. by 4-ft. planter (bottom right photo, facing page), we built three rolling planters for the deck. They're basically 2-ft. cubes made of pressure-treated frames, weather-proof casters, cedar siding and mahogany trim.

Our final job was to apply CWF wood preservative to the deck and planters. We applied two coats to the edges of the decking using a 1-gal. pressurized plastic garden sprayer with a narrow nozzle. A paint roller fitted with a 10-ft. extension handle was used to finish the surface of the deck. Bench tops were rubbed with progressively finer grades of steel wool between multiple applications of preservative. The residue was soaked up with a tack cloth.

Much to the clients' delight, the overall impact of the deck is striking. By day, the variegated color of the mahogany plays off the oiled cedar. By night, when the blue light from the spa constrasts with the raking yellow glow of the built-in lighting, the effect is almost dreamlike.                              □

*John Baldwin is a carpenter and writer in Greenwich, Conn. Photos by Bruce Greenlaw.*

# A Deck Built to Last

## Thoughtful details designed to beat the weather

by R. W. Missell

**W**e get a lot of pleasure from the birds and other wildlife that live near our house here in Virginia. So when we decided to build a deck, it was important that it provide a transition from our living space to theirs. The house backs into the woods, and we wanted a graceful deck that wouldn't obstruct the view. On many other decks I have seen, the assembly of railings and supporting posts masks the view with a clumsy grid of horizontal and vertical lines. But I also wanted to make sure that the deck would stay strong and solid for decades, so it features a number of details that are designed to ward off the destructive effects of the weather.

**The design stage**—I usually know how the parts of a finished project should fit together. But as a novice carpenter, I had to learn some hard lessons about the nitty-gritty of putting the parts in place. I found that it's less frustrating (and less expensive) to work problems out with a pencil rather than with a saw.

To prepare for this project, I worked up a scale drawing that included a plan view, elevation views and details. I also drew—and drew again—lots of pencil sketches to help me understand how each piece would be made and how each joint would fit together.

I had to balance code requirements and aesthetics. For example, our county's policy

requires 4x4 posts to support a deck railing. I knew these would look bulky, so my permit application included calculations showing my railing design (2x5 posts on 42-in. centers) to be stronger than required (4x4 posts on 60-in. centers). This data is found in tables entitled "Properties of Sections" in *The Wood Book* (Hatton—Brown Publishers, Inc., 610 S. McDonough St., Montgomery, Ala. 36197.

I wanted a deck pattern that would be decorative and that would visually break up the 30-ft. length of the deck. After looking at a variety of options, I decided to use a herringbone pattern made from 2x4 deck boards, run at a 45° angle. Making each "panel" 5 ft. wide meant that the longest deck board would be just over 7 ft. long, so I could start with standard 8-ft. boards. This minimized expensive waste.

Once I had the design squared away, it was easy to make up an accurate bill of materials. As a retail customer, I was able to negotiate only a relatively small discount on the initial bulk order. I knew I'd pay dearly for what I forgot—both in dollars and in lost time. On the other hand, liberal estimating could leave me with a lot of leftover materials that would be hard to store or to resell. In the end, my detailed material take-off was well worth the effort. On clean-up day, the only wood left over was a few treated 2x4s, which were quickly used in a landscaping project, and some 2x2s, which now serve as tomato stakes.

**Figuring a foundation—**With the planning complete and the lumber stacked in the garage, safely out of the sunlight to keep it from warping, construction could begin. One side of the deck would be supported by three separate posts. Two of the posts were supported on a concrete-block retaining wall, which had been built two years earlier with the future deck in mind. The rebar-reinforced blocks had been filled solid with concrete, and the footing was oversized in order to support the deck weight. When it came time for the deck, all I had to do was install anchor bolts for the posts (photo at right).

The third post had to be located in an existing sidewalk. I used a hammer drill fitted with a masonry bit to "perforate" a 12-in. square area of pavement, and then chipped the rest away with a hammer and chisel. Then I was able to excavate a pyramid-shaped cavity 24 in. deep, so that the finished footing would spread the load on undisturbed soil below the frost line. The hole was filled with concrete around wire-mesh reinforcement, and a J-bolt was inserted to anchor the post.

**Sizing structural elements—**A primary goal was to build a solid deck that would last at least 20 years and require an absolute minimum of maintenance. For this reason, I was conservative in sizing support members for strength and very careful at those points where wood rot could be expected. Although not required by our code, the galvanized steel bases I put under each post completely separate them from the concrete. A wood-to-concrete connection is one of the first places to rot, but the steel support provides a ½-in. air gap, thus ensuring good ventilation at the end grain of the post.

I decided that the posts would be 6x6s and the lengthwise support beam would be a pair of 2x12s. Local building code permitted 4x4 posts and a 2x8 beam, and upgrading these key components added only $55 to the cost of the project— a good long-term investment.

Where the support beam meets a post, the post was double-notched at the top in order to support a 2x12 on either side of a 1½-in. thick tongue. I sandwiched 2x12 scraps between the 2x12s of the support beam wherev-

The deck rests partially on a concrete-block wall that had been built some years before, but with the later construction of the deck in mind. The cores of the block had been reinforced and poured solid, so holes had to be hammer-drilled into the concrete for the anchor bolts (as shown in this photo), which were later grouted in place.

er a splice was required, and also at the ends of the beam. All of these components were through-bolted with ½-in. galvanized carriage bolts. The completed substructure formed a rigid, integral unit that helped to make the finished deck rock solid (photo adjacent page).

**Ledger details—**The support beam, located just beyond the lengthwise centerline of the deck, carries about two-thirds of the deck's weight. The ledger, however, is equally important from a structural standpoint because it ties the whole deck structure to the house as well as supports the remainder of the deck. Attaching the ledger board was a simple matter. I just removed the siding and used lag screws to attach the ledger board to the rim joist of the house.

The decking boards would have to be nailed to the ledger, and it was clear at the design stage that this wouldn't be easy. The 2x6 ledger provided only 1½ in. of nailing surface,

and even this was partially obstructed by the siding. As a result, the deck boards would have only an inch of bearing surface on the ledger. This connection would certainly be a weak link in the structure. And beside that, I didn't like the idea of driving all those nails so close to the aluminum siding, a material easily scarred by errant hammer blows. Instead, I nailed in a 2x4 flat to the top of the ledger (bottom photo, next page). The flat side of the 2x4 solidly supports the ends of the deck boards, and each of the 2x8 joists hanging from the ledger would have to be notched to clear the nailer. But the nailer was worth the trouble, because it reduced the likelihood of split end grain in the decking and also kept hammer dings in the siding to a minimum.

Once the siding had been removed to install the ledger, the house's main structural system was exposed to the elements. Rain and snow would reach into the voids and joints at the end of each deck board, and it

In order to shield fasteners from the weather, pressure-treated plywood was first screwed to the tops of the 2x railing supports. Screws were then run in through the plywood to the 2x railing cap to fasten the cap in place. The plywood also acts as a splice plate where a joint is required between supports.

was vital to protect the untreated rim joist from this moisture. The solution was a simple flashing strip, which I slipped under the siding and extended over the top of the nailer strip to form a drip edge well away from the rim joist (photo bottom left). Admittedly, this flashing was perforated with nail holes when I nailed in the deck boards, but the joint between the nailer and the rim joist will still be protected. Next time around, though, I think I'd caulk beneath the flashing, too.

**Joist details**—For a given load, the correct joist size depends on how far they span and on how closely they're spaced. But figuring the proper spacing wasn't so easy for this deck. First, the diagonal pattern of the decking boards meant that each board would span a greater distance than the distance between joists. Second, a joist would be needed wherever the decking boards changed direction (every 5 ft.). I settled on a spacing of roughly 14 in. o. c. for the joists, resulting in a clear span of roughly 19 in. for the longest deck board. With this reduction in spacing, 2x8 joists were a conservative choice.

Along each line where decking boards changed direction, I doubled the joists, and spaced them apart with scraps of 2x. This provides a full 1½-in. support area at the end of each deck board, and nails can be sunk well away from the ends of the boards, which virtually eliminates end splits. It also provides an open space between joists so that a single pass with a circular saw trims all deck boards evenly. Most important, however, it ensures that the butt ends of the deck boards will be well-drained and ventilated. Tightly made joints directly over a joist will eventually rot as water collects around them. The doubled joists are supported by heavy-duty joist hangers (United Steel Products Co., 703 Rogers Drive, Montgomery, Minn. 56069).

**Blending the design elements**—All construction details to this point were designed to increase the strength and durability of the deck. But after the joists were in place, appearance became a major consideration. The deck's dominant visual element is the sanded and rounded 2x surfaces. All the 2x4 deck planks and exposed boards in the railing and

The key element of any deck is the support system of posts, beams and joists. In this project, posts were notched to support a beam built from doubled 2x12s. The post extends between the 2xs as a tongue and spaces them apart. Bench supports were solidly bolted to the substructure (above). Below, deck joists hang from a ledger lag-bolted to the house. The black joist hanger on the left is wide enough for doubled joists spaced apart with 2x blocking. Flashing slipped under the siding directs water away from the house. Just visible beneath the flashing is a 2x4, nailed flat to the top of the ledger that provides additional support to the deck boards.

The completed deck includes herringbone decking and a privacy screen at one end. Note the rounded front portion of the benches, a detail designed to increase comfort. Edges on the benches and on the railing were sanded smooth.

benches were belt-sanded smooth, and the edges were rounded over with a router and a ½-in. radius corner rounding bit. For assembly-line efficiency, I set up a separate work area for these operations, including a bench vise, sawhorse supports and power cords.

**The railing system**—Several conflicting design goals affected the construction details for the railing. First, I wanted the view from a seated position on the deck to be as unobstructed as possible. My intent was for people on the deck to feel themselves a part of the adjacent woodland, and not feel fenced out by the deck rail. The final design uses evenly spaced, slender horizontal and vertical lines to create a see-through effect, much like a venetian blind (photo above). A key factor here is the use of 2x5 posts turned edgewise to the deck to minimize their mass, and edgewise 2x stock for railing and barrier slats. The county safety code requires a 36-in. rail height with openings between railing and deck held to a maximum of 6 in., but the thin profile of the edgewise stock allowed me to meet this requirement with a minimum of visual impact.

Besides keeping a slim silhouette when viewed from a chair on the deck, the railing was designed to offer a broad, flat area for flower pots, serving plates and cold drinks. To achieve this, I used a 2x8 with rounded edges for the railing cap and a 1x apron for trim. Each joint in the rail cap was mitered across the thickness of the cap and carefully fitted to provide a smooth, unbroken surface.

The heart of the railing system is the 4½-in. wide strip of treated ½-in. plywood that ties all the pieces of the cap together (photo top right, facing page). Installation wasn't too difficult. After ripping the plywood to width, I screwed lengths of it to the tops of the posts. Then I cut the 2x8 rail caps to length and attached them to the plywood with 2-in. galvanized woodscrews run in from beneath

Installing benches single-handed can be tough, but the liberal use of clamps can speed the job. The bench backs and seats were angled slightly to make them more comfortable.

through the plywood. I ripped the trim pieces from 1x stock and rounded the edges before nailing them to the underside of the railing. As a result, no nails are visible from above, and no plugs are required in the rail. Equally important, the plywood acts as a splice plate to keep butt joints tight and to reduce warping and cupping of the rail cap boards.

**Bench building**—I cut the bench supports from 2x8 stock, wanting to make sure they would be strongest where each support is bolted to the deck frame. I didn't want them to look clumsy, so each support is tapered to a dimension of 1½ in. by 4½ in. where it joins the railing (photo above). The bench supports were installed at an angle of 15° so that the seat back would be comfortable. The seat rails

were cut from 2x6s and mounted with 5° of tilt. Combined with the rolled edge at the front of each bench, these angles result in very comfortable seating. I used galvanized ½-in. carriage bolts to tie everything together.

**Privacy screen**—In order to block the view from a neighbor's yard, I added a privacy screen at one end of the deck (photo top). The screen was built in the same way as the rail system, but with more horizontal slats—these were spaced ½ in. apart. I added an extra support near the diagonal section of railing in order to support the ends of the slats. □

*R. W. Missell is an electrical engineer with 20 years of experience in the operation and maintenance of Air Force facilities.*

# Tassajara Makeover

## Sturdy joinery and durable materials form a new deck and bridge

by Gene DeSmidt

Fifteen years ago, during a West Coast trek, I walked into Tassajara Hot Springs through massive wood gates hinged to a log-framed, Japanese-style entry. The entry's concrete and stone foundation seemed to grow right out of the ground. Huge boulders were strewn about, each appearing to be purposefully placed. Surrounded by native pine, oak and sycamore, they offered a serene welcome to a weary traveler.

Since then, I've returned to Tassajara dozens of times to orchestrate a variety of construction projects as a breather from my hectic contracting business. This article will focus on one of my favorites: the rebuilding of a wood bridge that spans a creek and provides access to Tassajara's 110° hot springs, as well as the rebuilding of an accompanying deck. The project called for the construction of nifty details such as mortised-and-tenoned railings, bamboo-covered screens and gable-roof entries (photo above). Everything was built to endure high water, damp, cold winters and bone-dry summers, and called for a degree of craftsmanship befitting the location. Of course,

that meant living in this remote paradise for the duration of the job, or about six months. That was a sacrifice I could live with.

**Bridge building**—Tassajara is a Zen Buddhist monastery/hot springs resort, known in part for its vegetarian cooking and books on bread, as well as for its attraction to professionals and craftspersons in need of a respite. Located in a remote and rugged canyon about 30 miles east of Big Sur, it's reached by driving over a steep, bumpy, 15-mile-long dirt road.

Our daily routine began with bells and breakfast and a morning meeting not unlike what might have occurred in an ancient craft guild, or even in a modern small-town factory. Then we went to work.

Because the existing bridge was a safety hazard, we started there. Built in 1969 as a temporary replacement, its backbone was a pair of 35-ft. long girders, laminated on site by bending Douglas fir 1x4s around a tree and nailing them together. Untreated, the girders had rotted beyond repair. The new bridge

is supported by a pair of laminated, pressure-treated, 8x18 Douglas fir girders, cambered 2 ft. for a traditional appearance. They were designed by Paul Discoe of Joinery Structures in Oakland, engineered by Geoffrey Barrett from Mill Valley and custom-made by Barnes Lumber, a truss manufacturer in Cloverdale. The girders were trucked to Tassajara on a 24-ft. flatbed.

We began demolition and construction in late summer, when the river was low. We started by cutting down the old bridge with chainsaws, letting it sit temporarily beneath the location of the new bridge to serve as a scaffold. Then we built a simple dolly to support the leading ends of the girders, screwed 2x handles to their trailing ends and (with the help of 10 people) rolled the unwieldy monsters across the old bridge, one at a time.

The next step was to bolt the new girders 4 ft. o. c. to a new redwood "trestle" on the bathhouse side of the river and to an existing concrete foundation on the opposite side. The trestle consists of two built-up 6½-in. by 7½-

Photo: Dan Howe

From *Fine Homebuilding* (June 1991) 68:70-73

**Bridge anatomy**

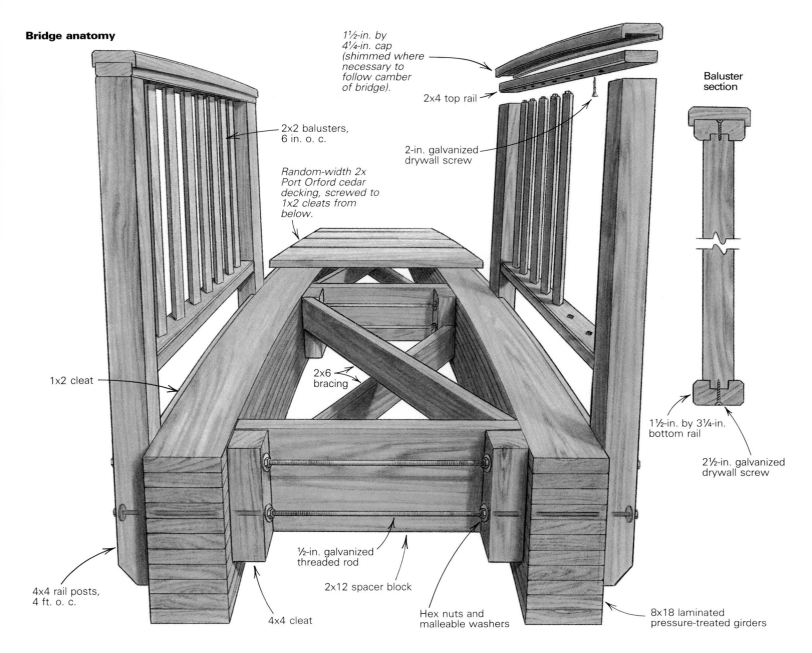

1½-in. by 4¼-in. cap (shimmed where necessary to follow camber of bridge).

2x4 top rail

2-in. galvanized drywall screw

2x2 balusters, 6 in. o. c.

Random-width 2x Port Orford cedar decking, screwed to 1x2 cleats from below.

1x2 cleat

2x6 bracing

Baluster section

1½-in. by 3¼-in. bottom rail

2½-in. galvanized drywall screw

½-in. galvanized threaded rod

2x12 spacer block

4x4 rail posts, 4 ft. o. c.

4x4 cleat

Hex nuts and malleable washers

8x18 laminated pressure-treated girders

---

in. posts capped with a 6x14 beam and braced diagonally with a pair of 2x6s. The posts are planted in a reinforced-concrete foundation to which they're pinned horizontally with ½-in. stainless-steel threaded rod.

With the girders anchored, the next step was to tie them together and to fasten 4x4 railing posts to their outside faces on 4-ft. centers. We accomplished both at once through the use of ½-in. galvanized threaded rod.

The technique was simple. First, we marked the locations of the railing posts on the girders and tacked 4x4 cleats vertically to the inside faces of the girders (drawing above), directly opposite the layout marks. Next we nailed 40-in. long, 2x12 spacer blocks to the 4x4 cleats. Posts were then installed in opposing pairs by running a single length of threaded rod through both posts, both girders and the two corresponding 4x4 cleats. We used hex nuts and malleable washers to simultaneously tighten the posts against the girders and the girders against the 2x12 spacer blocks (malleable washers have a larger diameter, for

more bearing, than standard washers, and we liked the way they look). A second pair of nuts and washers serves to back-tighten the posts to the girders. Finally, to prevent the bridge from swaying, we installed 2x6 diagonal bracing in each 4-ft. section.

**Fitting the railings**—Between the posts, the bridge railings consist of 2x2 balusters spaced 6 in. o. c. and mortised into 2x4 top rails and 1½-in. by 3¼-in. bottom rails. The top rails sit atop the posts, while the bottom rails are mortised into them. We cut the mortises, using a router equipped with a straight bit, and the tenons with Japanese handsaws. For a decorative effect, we turned the balusters 45° to the rails (we called them "diamond pickets").

Up top, we dadoed a 1½-in. by 4¼-in. railing cap to slip over the top rails. The ½-in. deep dado rendered the cap flexible enough to follow the curvature of the bridge, disguising the straight lengths of the top rails. The dadoes were cut using a Makita groover, a handy tool that resembles a power plane

equipped with dado knives (unfortunately, Makita doesn't sell them in the U. S.).

We used Port Orford cedar for the railings (and for most of the other woodwork on the site). It's a fragrant, rot-resistant, exceptionally stable wood that's used for everything from closet linings to canoe paddles. The wood is native to southern Oregon and northern California, but it's in such demand in Japan that it's becoming relatively scarce and expensive in the U. S. We bought ours from a small mill in Oregon and had it trucked directly to the site.

We prefabricated the balustrades in sections by screwing the top and bottom rails to the balusters with 2½-in. long galvanized drywall screws. That made it easy to install the sections between posts as we worked our way across the bridge. We screwed the top rails to the tops of the posts with drywall screws and, once all the sections were installed, fastened the caps from underneath the top rails using 2-in. long galvanized drywall screws, shimming the caps where necessary to follow the camber. Finally, we chamfered

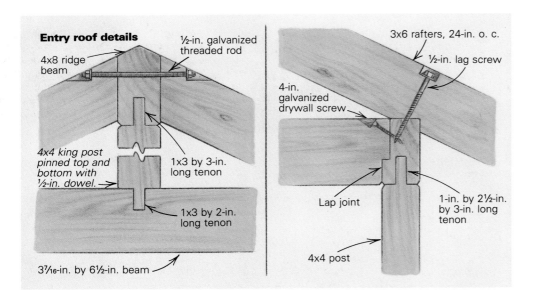

**Entry roof details**

4x8 ridge beam

½-in. galvanized threaded rod

4x4 king post pinned top and bottom with ½-in. dowel.

1x3 by 3-in. long tenon

1x3 by 2-in. long tenon

3⁷⁄₁₆-in. by 6½-in. beam

3x6 rafters, 24-in. o. c.

½-in. lag screw

4-in. galvanized drywall screw

Lap joint

1-in. by 2½-in. by 3-in. long tenon

4x4 post

and planed the caps to a glasslike finish using Japanese hand planes.

With the railings completed, we screwed 1x2 cleats to the top outside edge of the girders. Then we screwed random-width, 2x cedar decking to the cleats from below so that no screws would show on top. The new bridge was finished with Penofin (Performance Coatings, Inc., P. O. Box 1569, Ukiah, Calif. 95482; 707-462-3023) before we completed the demolition of the old one. We chose Penofin because we liked the golden tone it imparted to the cedar.

**Fixing the underpinnings**—A year before replacing the bridge, we had excavated and poured a continuous 24-in. wide by 12-in. deep concrete grade beam in line with and beneath the existing concrete deck columns. The grade beams were designed to prevent the columns

Photo: Charles Miller

Framing for the men's entry consists of 4x4 posts mortised to the deck and to a 4x top plate that supports the roof. The entry's interior is finished with 1x6 Western red cedar and holds a simple bench made of 3x6 Port Orford cedar.

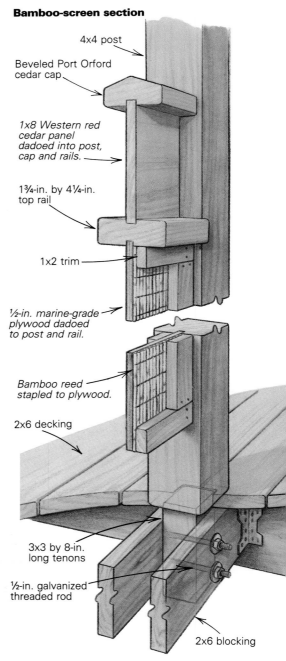

**Bamboo-screen section**

4x4 post

Beveled Port Orford cedar cap

1x8 Western red cedar panel dadoed into post, cap and rails.

1¾-in. by 4¼-in. top rail

1x2 trim

½-in. marine-grade plywood dadoed to post and rail.

Bamboo reed stapled to plywood.

2x6 decking

3x3 by 8-in. long tenons

½-in. galvanized threaded rod

2x6 blocking

from being undercut by the river's current and to buttress them against careening boulders.

Now that the new bridge was completed, it was time to tear out the old deck, beef up the battered columns and build a new deck. We enlarged the 8x8 columns to 12x12s by wrapping a cage of ½-in. rebar around each one, enclosing the cages with forms built out of ¾-in. plywood and 2x4s, and then filling the forms with concrete. The columns were capped with stainless-steel brackets (custom-made by Stoltz Metals in Oakland) secured with anchor bolts set into the concrete.

With the columns repaired, we bolted a series of 6x12 pressure-treated girders to the stainless-steel brackets, parallel to the river. Then we installed 4x8 pressure-treated joists perpendicular to the girders on 3-ft. centers, sufficient to support the covered benches, privacy screens and other structures that would eventually sit on top of the deck.

The joists are supported at the girders by joist hangers, and at their opposite ends by pockets cold-chiseled out of the concrete foundation supporting an adjacent bathhouse. We installed the joists 1½ in. below the tops of the girders so that the decking would be flush with the girders and nearly flush with the tile floors in the bathhouse.

The decking is 2x6 Port Orford cedar, fastened with 3½-in. silicon bronze, ring-shank boat nails. These nails, which cost about $6/lb., are very durable and hold so tenaciously that they're almost impossible to pull without destroying the wood in the process. I bought ours from R. J. Leahy & Co. (486 8th St., San Francisco, Calif. 94103; 415-861-7161).

To widen the walk at the front of the deck, we also installed one deck board on the outboard side of the girders. This was accomplished by nailing 4x blocks to the front of the girders on 3-ft. centers, nailing a 2x12 cedar skirtboard to the blocking, and installing the deck board over the blocking and skirtboard. The skirtboard conceals the face of the girders.

**Bamboo for privacy—**The design of the deck top, including the layout of the railings, bamboo-covered privacy screens and covered structures, is primarily the handiwork of Berkeley architect Mui Ho. Many of the structural and visual details, however, were worked out on site by me and my crew.

The deck railings mimic those of the bridge, except for a few details. The top and bottom rails are both mortised to the posts and are 3 in. wide; the rail caps aren't dadoed on the bottoms; and the rail posts are fastened with ½-in. lag screws and malleable washers to the skirtboard and girders.

The bamboo-covered screens are framed with 4x4 posts spaced 4 ft. o. c. and capped with 1¾-in. by 4¼-in. top rails, both dadoed to receive ½-in. marine-grade plywood panels (bottom drawing, facing page). The screens at the front of the deck were installed by lag-screwing the posts to the skirtboard, same as the railing posts. We wanted the rest of the screens to appear as though they're freestanding, with the posts simply sitting on the deck. We accomplished this by mortising a series of 2½-in. square holes in the decking to accept 8-in. long tenons cut on the bottoms of the posts. Every third tenon was fastened directly to a deck joist with two lengths of ½-in. galvanized threaded rod. The remaining tenons were sandwiched between and bolted to a pair of 2x6 blocks installed perpendicular to the joists. We cut the mortises using a jigsaw, and the tenons by making several passes with a circular trim saw and knocking out the waste with a hammer and a wood block. The tenons were cleaned up using a razor-sharp slick.

Before installing the plywood panels, we soaked them with a water-repellent preservative and painted the interior faces (facing the bathhouse) with an off-white exterior latex paint. Then we slipped them into their dadoes and toenailed them to the posts through their exterior sides. We stapled bamboo reed (bamboo strands woven together with galvanized wire) over the exterior sides of the panels, trimmed the exterior sides with 1x2s installed with a nail gun, and applied Penofin to the cedar.

As soon as we installed a section of the screen, we found that its 6-ft. height was too short to block the view fully from the apex of the bridge. Because post height was limited by the length of our stock, our solution was to dado 1x8 Western red cedar panels into the top rails and to cap the panels with dadoed lengths of 2-in. by 3-in. Port Orford cedar beveled to a peak on top (bottom drawing, facing page). Every 8 ft. to 10 ft., we lag-screwed a short, dadoed 3x3 post to the top rail directly over a 4x4 post to support the ends of the 1x8 panels. The solution worked and added a nice decorative touch to the tops of the screens.

**The old steps consisted of slippery river stones that were a hazard to barefoot bathers. The steps were replaced by new ones carved on site with a pneumatic hammer out of locally retrieved sandstone. The natural texture of the stone is virtually slip-proof.**

**Posts and beams—**In all, the deck supports five gable-roofed structures: one highlighting the entrance to the bathhouse complex; two serving as entries to men's and women's bathing areas; one each for a storage shed and the women's shower. The shower and storage shed were stick-framed and stuccoed, while the remaining structures were framed with Port Orford cedar posts and beams.

Each post-and-beam structure consists of four 4x4 posts fastened to the deck in the same way as the screen posts. Above, the posts are mortised into a continuous 4x beam that supports the roof (top drawings, facing page). We lap-jointed and screwed the beams together at the corners, allowing the two beams that support the rafters to cantilever past the posts to carry the roof overhangs.

The 3x6 rafters were cut like typical common rafters. To install them, we toenailed them on 2-ft. centers to the ridge beam through the tops. Then we drilled a ⁹⁄₁₆-in. hole clear through each rafter pair and the ridge. That allowed us to insert a length of ½-in. threaded rod through each hole and to tighten the rafters against the ridge with hex nuts and washers. The nuts and washers were countersunk to prevent them from protruding above the rafters. We screwed the bottom of each rafter to the supporting beams with a ½-in. lag screw.

This system of attachment strengthened the roof framing markedly even before we nailed down the 1x6 Western red cedar roof decking. Once we installed the decking, we finished the roof with 15-lb. felt and 30-year fiberglass shingles. For a decorative touch, we chamfered most of the edges of the framing members with a router.

The walls of the men's and women's entries are of the same construction as the bamboo screens, except that instead of painting the interior sides of the ½-in. plywood, we covered them with Western red cedar 1x6s, chamfered on the edges and fastened horizontally. The Neapolitan-colored Western red cedar contrasts nicely with the Port Orford cedar framing. The entries were finished with Penofin.

**Chiseling the steps—**To reach the ancient steam rooms from the deck, bathers used to have to slip and stumble over an awkward set of river-rock steps. Feeling strongly that this invited a broken ankle or worse, I commissioned San Francisco sculptor Robert Gove to carve a set of solid sandstone steps to replace the old ones (photo left).

The new stone steps not only look good, but their naturally rough texture makes them virtually slip-proof (a concrete-filled copper handrail was installed later).

Setting the last step provided the finishing touch on the bathhouse project. The now completed bridge, deck and stairs were ready for the pleasure of the Tassajara guests and resident bathers. □

*Gene DeSmidt is a contractor in Oakland, California.*

# Railing Against the Elements

## With careful detailing and proper flashing, exterior woodwork should last for decades

by Scott McBride

**Mushrooming problems.** When fungi sprout from railings, it's a rotten sign. Pictured here is a close-up of the railing the author was hired to replace.

The moist climate of New York's Hudson Valley isn't exactly ideal for carpenters like myself. We have muggy summers, slushy winters and three months of rain in between. As we struggle through weeks of unremitting precipitation with fogged-up levels and wet chalk lines, there is but one consolation: come April, a billion fungal spores will bloom, reaching into every water-logged mudsill, fascia and doorstep. That means a guaranteed crop of rot-repair jobs in the coming season.

Of all the woodwork exposed to the elements, none is so vulnerable as the white pine porch railing. With the right combination of faulty detailing and wind-driven rain, a railing can be reduced to shredded wheat in about eight years. I typically rebuild several of these railings each year. In this article I'll describe one such project (photo facing page).

**Getting organized**—A railing around the flat roof of a garage had rotted out (photo above), and the owner asked me to build a new one. This wasn't a deck that was used, so the railing was decorative rather than functional, and one of my worries was installing the new railing without making the roof leak (more about that later). On a house across town, my customer had spotted a railing that he liked and asked if I could reproduce it. I said I could and took down the address.

Before leaving the job site, I made a list of the rail sections I would be replacing, their lengths and the number of posts I would need. Later that afternoon I found the house with the railing my customer liked, strolled up the walk and began jotting down measurements. The family dog objected strenuously to my presence, but no one called the police.

Back at the shop I drew a full-scale section of the railing and a partial elevation showing the repeating elements. The next bit of work was to make layout sticks (or rods) showing the baluster spacing for the different rail sections (see the sidebar on p. 52). This would tell me the exact number of partial- and full-length balusters that I would need.

**The right wood**—I'm fortunate to have a good supplier who specializes in boat lumber. He carries premium grades of redwood, cedar, cypress and Honduras mahogany, all of which resist decay well. I used cypress for the rails and balusters because it's less expensive and

because the rough stock is a little thicker than the others. Cypress is mostly flatsawn from small trees, though, and the grain tends to lift if the wood isn't painted immediately.

Much of the Western red cedar at this yard is vertical grain—the annual rings run perpendicular to the face—and hence inherently stable. I typically use it for wider pieces where cupping could be a problem. Square caps on posts fit this description because they are so short in length. They should always be made from vertical-grain material or they'll curl in the sun like potato chips.

For this job I bought 2-in. thick cypress for the rails and balusters, 5/4 red cedar for the rail caps and post caps, and 1-in. cedar boards for the box newel posts.

**Bevels and birds' mouths**—After cutting the lumber into rough lengths with a circular saw, I jointed, ripped and thickness-planed the pieces to finish dimensions. Using a table saw, I beveled the top, middle and bottom rail caps at 15° (drawing facing page). Besides looking nice, the bevels keep water from sitting on what would otherwise be level surfaces.

I had to cut a bird's mouth on the bottom of

From *Fine Homebuilding* (October 1991) 70:68-71

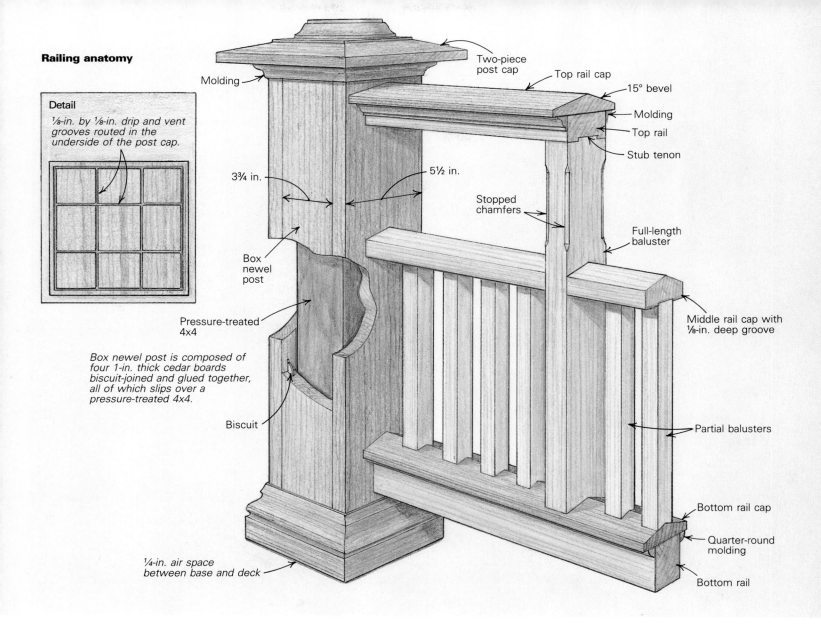

**Railing anatomy**

Molding

Two-piece post cap

Top rail cap

15° bevel

Molding

Top rail

Stub tenon

**Detail**

⅛-in. by ⅛-in. drip and vent grooves routed in the underside of the post cap.

3¾ in.

5½ in.

Stopped chamfers

Full-length baluster

Box newel post

Pressure-treated 4x4

Middle rail cap with ⅛-in. deep groove

*Box newel post is composed of four 1-in. thick cedar boards biscuit-joined and glued together, all of which slips over a pressure-treated 4x4.*

Biscuit

Partial balusters

Bottom rail cap

Quarter-round molding

¼-in. air space between base and deck

Bottom rail

each partial baluster so that the baluster would fit over the beveled cap on the bottom rail. Rather than make each bird's mouth individually, I first crosscut my 2x6 baluster stock to finished length and jointed one edge. I then ripped just enough off the opposite edge to make it parallel to the jointed edge. With the blade on my radial-arm saw raised off the table and tilted 15°, I made an angled crosscut halfway through the thickness of the baluster stock. Flipping the piece, I made the same cut from the opposite face. Individual balusters were then ripped from the 2x6, with each bird's mouth already formed.

I used the same setup to cut birds' mouths in the full-length balusters, but cut them one at a time because they were wider. These uprights also have decorative stopped-chamfers routed into them. The chamfer bounces light smack into your eye in a most appealing way.

Next I set up the dado cutterhead on my table saw. I ploughed a shallow groove for the partial balusters in the underside of the middle rail, and another to receive the full-length balusters in the underside of the top rail. Although these grooves are a mere ⅛ in. deep, they made assembling the railing much easier

**Decorative railings.** *The original railing around the deck over the garage rotted out prematurely, so the owner of the house commissioned a new railing of a slightly different design—one copied from a house across town. Carefully detailed of cypress and cedar, the new railing should last a long time. Photo by Kevin Ireton.*

and ensured positive alignment of the vertical members. The tops of the partial balusters are housed completely in the groove, but the tops of the full-length balusters also have stub-tenons, cut on the radial-arm saw.

**Box newel posts—**Each box newel post is a two-part affair. I cut rough posts from pressure-treated yellow pine 4x4s and outfitted them with soldered-copper base flashing (top drawing, next page). This flashing would later be heat-welded to the new roof surface (more on that in a minute). A finished cedar box newel post would slip over the rough post and receive the railings.

Because the sides of the box newels were to be butt-joined, two sides were left their full 5½-in. width, and the other two were ripped down to 3¾ in. so that the finished post would be square in cross section. I saved the rippings to make the quarter-round molding that's under the bottom rail cap.

The box newels were glued up with a generous helping of resorcinol to keep out moisture. Glue also does a better job of keeping the corner joints from opening up than nails do. To align the sides of the box newels dur-

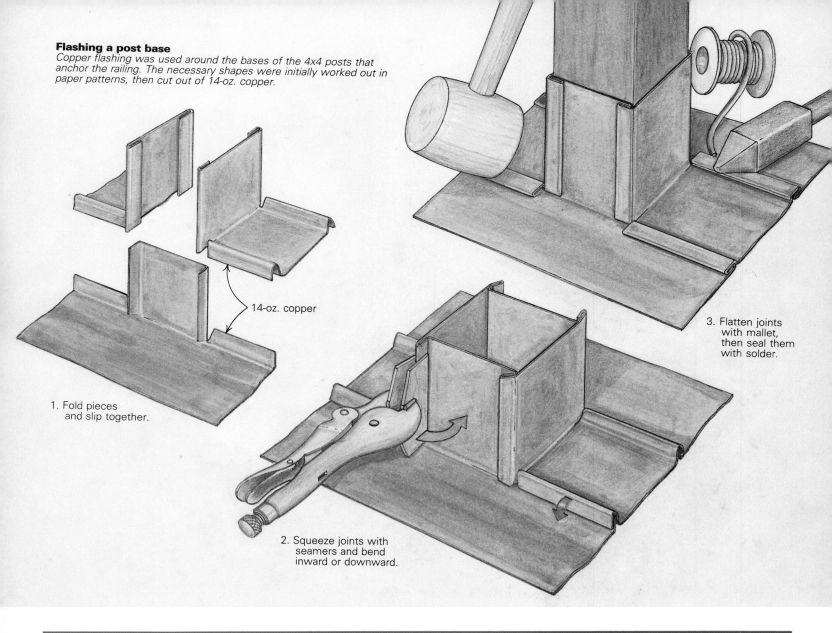

**Flashing a post base**
*Copper flashing was used around the bases of the 4x4 posts that anchor the railing. The necessary shapes were initially worked out in paper patterns, then cut out of 14-oz. copper.*

14-oz. copper

1. Fold pieces and slip together.

2. Squeeze joints with seamers and bend inward or downward.

3. Flatten joints with mallet, then seal them with solder.

## Spacing balusters

Spacing balusters correctly is a simple trick, but it's surprising how many carpenters are stumped by it. The object is to have equal spaces between each baluster and between balusters and posts.

Suppose your railing is $71\frac{5}{8}$ in. long., your balusters are $1\frac{1}{2}$ in. square, and you want a spacing of 4 in. o. c.—that is, $1\frac{1}{2}$ in. of wood, $2\frac{1}{2}$ in. of air, $1\frac{1}{2}$ in. of wood, and so on.

Begin the layout from the post on the left, by tentatively laying off a $2\frac{1}{2}$-in. space ending at point A in the drawing below. From there you stretch a tape and see how close you come to the post on the right with a multiple of 4 in. In this case 68 in. (a multiple of 4) brings us to point B, which is $1\frac{1}{8}$ in. from the right-hand post. Forget about this remainder for the moment.

Dividing 68 by 4 tells you that you will have 17 intervals containing one baluster and one space, plus the extra space you laid out at the start. That makes 17 balusters and 18 spaces. Now you're going to lay out the combined dimensions of all 17 balusters ($17 \times 1\frac{1}{2}$ in. $= 25\frac{1}{2}$ in.) from the left-hand post. That brings you to point C. From there you're goint to lay off the combined dimensions of all the spaces, bringing you to point B—close, but not close enough. To land directly at the post, divide the remaining $1\frac{1}{8}$-in. by 18 (the number of spaces) and add the quotient to each space. How fortunate $1\frac{1}{8} \div 18 = \frac{1}{16}$. That gives you a nice, neat adjusted dimension for the space of $2\frac{9}{16}$ in.

If the numbers don't divide evenly, I'll use a pair of spring dividers to find the exact space dimension by trial and error, stepping off the distance from Point C to the post. When I find the right setting, I lay off the first space. Then I add this dimension to the baluster width, reset the spring dividers to the sum distance, and step off the actual spacing. This method avoids the accumulated error that happens when using a ruler and pencil, not to mention all that excruciating arithmetic. —S. M.

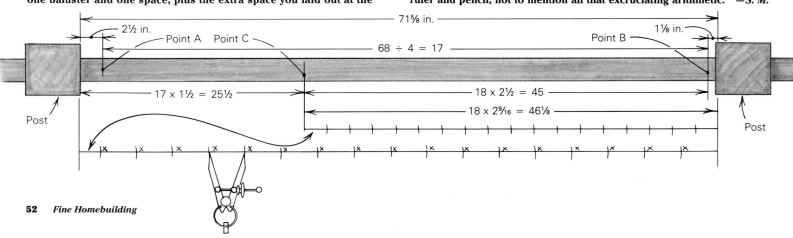

$71\frac{5}{8}$ in.

$2\frac{1}{2}$ in.

Point A    Point C

$68 \div 4 = 17$

$1\frac{1}{8}$ in.

Point B

$17 \times 1\frac{1}{2} = 25\frac{1}{2}$

$18 \times 2\frac{1}{2} = 45$

$18 \times 2\frac{9}{16} = 46\frac{1}{8}$

Post

Post

ing glue-up, I used three biscuit joints along the length of each side.

The last parts to be fabricated were the post caps. The square lower part of the cap is a shallow truncated pyramid, produced by making four consecutive bevel cuts on a table saw. I made the round upper part on a shaper. The two parts were glued together with the grain of each parallel to the other, so they would expand and contract in unison.

To reduce the amount of water running down the face of the post, I cut a drip groove around the underside of the cap with a ⅛-in. veiner bit mounted in a router table. Using the same bit, I also routed a series of ventilation grooves into the underside of the cap in a tic-tac-toe pattern (detail drawing, p. 51). They allow air taken in at the bottom between the rough post and the box newels to escape at the top without letting in rain.

**Assembling the rail sections**—I assembled the rail sections in the shop where I could count on dry weather and warm temperatures. The first step was to face-nail the bottom rail cap to the bottom rail. I placed the nails so that they would be covered subsequently by the ends of the balusters.

I toenailed the balusters in place with 4d galvanized finish nails. This was easy to do because the bevel and bird's-mouth joinery prevented the balusters from skidding around as I drove the nails.

The middle rail was cut into segments to fit between the full-length balusters. I took the lengths of the segments, along with the spacing for the partial balusters, directly from the layout on the bottom rail cap. Every other segment of the middle rail could be attached with 3-in. galvanized screws through the uprights. The intervening segments were toenailed with 8d galvanized finish nails. Then I face-nailed through the middle rail down onto the tops of the partial balusters. The top rail was screwed down onto the full-length balusters. I left the top-rail cap loose so that it could be trimmed on site for a tight fit between the newels.

Still in the shop, I caulked all the components with a paintable silicone caulk, primed the wood and then painted it with a good-quality latex house paint. With a truckload of completed rail sections and posts I headed for the job site with my crew.

**Installation**—My roofing contractor had replaced the existing 90-lb. rolled roof over the garage with a single-ply modified-bitumen roof. One advantage of modified bitumen is that repairs and alterations can be heat-welded into the membrane long after the initial installation. This meant I didn't have to coordinate my schedule with that of the roofer to fuse the copper base flashings to the new roof. Flashing strips of the bitumen were melted on top of the copper flange (photo right), providing two layers of protection (including the copper) around the base of the post and a "through-flashed" layer of roofing beneath the post.

The railing is U-shaped in plan, and I anchored the two ends into the house. However,

**Fused flashings.** One of the advantages of a single-ply modified-bitumen roof is that repairs and alterations can be heat-welded to the membrane long after the initial installation. Here, flashing strips of bitumen, which cover the copper base flashings around the 4x4 posts, are being fused to the roof membrane.

I didn't want to penetrate the roof membrane with framing or fasteners, so the newel posts are only attached to the deck by way of their flashings. Although this method of attaching the posts provides superb weather protection, the intermediate posts are a bit wobbly. This was okay for this particular deck because the railing is strictly decorative. Where a roof deck is subject to heavy use, the posts should be securely anchored to the framing.

With the rough posts in place, installation of the railing was straightforward. Box newels were slipped over the 4x4s and roughly plumbed. We stretched a line between corner newels and shimmed the intermediate newels up to the line. All box newels were held up off the deck at least ¼ in. to allow ventilation. The difference between the rough post dimension (3½ in. by 3½ in.) and the internal dimension of the box newels (3¾ in. by 3¾ in.) allowed for some adjustment. When the newels were just where we wanted them, we simply nailed through the shims and into the rough posts.

In some cases rail sections had to be trimmed. When the fit was good, top, middle and bottom rails were snugged up to the posts with galvanized screws. All that remained was to glue cypress plugs into the counterbored holes and shave them flush. □

---

*Scott McBride is a builder in Sperryville, Va., and a contributing editor of* Fine Homebuilding. *Photos by author except where noted.*

# Building a Winding Outdoor Stair

## Fitting stairs to a complicated site called for an accurate sketch and some job-site improvising

by Thor Matteson

A couple of summers ago, I was helping my father build a house in El Portal, California, when some neighbors, Ron and Liz Skelton, asked if I'd be interested in building a deck for them. They offered to *pay* me for my work; my father understood the lure of fortune and fame and encouraged me to accept the job, which I did.

The Skeltons wanted a redwood deck to replace a sagging fir assembly built over a rock garden. That part of the job would be tricky enough, but there was more. Ron is the local mechanic, and he had no stairs leading from the house down to his shop located one level below. After several years of skidding down the bank to work, he was ready for a stairway. Ron described to me where he and Liz envisioned the stairs—"starting about here, winding down between these rocks and ending up at the slab in front of the shop," he narrated while scampering down the bank around several boulders. Suddenly I felt slightly dizzy. I was barely getting the hang of *straight* stairs, let alone double-inflected variable-radius curved stairs. Nevertheless, I forged ahead with the deck and extended a landing from which to start the stairs.

Liz and Ron wanted a sturdy-looking stair, so we settled on 42-in. wide (approximately) open-string stairs with 2x12 stringers and 2x6 risers, both of redwood, and 4x redwood treads (photo facing page). Four-in. thick redwood can be tough to find, and we needed 4xs for the deck framing, too. Luckily, a local contractor gave Ron a tip on where to get it, and the supplier sent the most stunning redwood I'll probably ever see—vertical grain, surfaced full-dimension 4x6s and 4x8s, 16 to 20 ft. long with hardly a defect in them. I saved the nicest stuff for the treads and sentenced the rest to life beneath the decking.

**Measuring and sketching**—When building an unconventional stair such as this one, it's a good idea to start with a drawing (drawing facing page). I mapped out a plan view first, scaled ½ in. to the foot and identified the critical elevations. That called for a few hours scaling the hillside with a tape measure, a plumb bob and a builder's level.

To figure out the total rise of the stair, I perched the builder's level near the top-land-ing location and shot the elevations of both the top landing and the shop slab (the bottom slab of the stair would butt up level to the shop slab). Then I subtracted the top elevation from the bottom and came up with a total rise of 127 in. I don't have a leveling rod to use with my level, but a tape measure attached to a straight 2x4 served me just as well.

Next, using framing members of the new deck as reference points, I measured horizontally and dropped a plumb bob to locate the relative positions of the three boulders around which the stairs would wind, plus the location of the bottom landing. I used all of these measurements that evening to draft a plan view of the entire project. The path the stair would have to take was now obvious, and I penciled it into the drawing.

Now I drew in a line which I thought best represented the line of travel someone was likely to take in traversing the stairs, picturing myself swinging around the end post on the deck railing and trotting down the steps, brushing past the boulders while taking the shortest route to the shop. Then I laid out the stair so that the rise and run of each step would be consistent along that line of travel. I walked my architect's scale along the curved path line, and I decreed the scaled length of the path line to be the total run of the stairs— 20 ft. even (by coincidence). I converted this figure to inches and used this number, along with the 127-in. rise to figure out the rise and run of each step. The rule of thumb for stairs states that rise plus run should be between 17 and 18 in. and that rise times run should equal about 75 in. I decided on a rise of 6¾64 in. and a run of 12 in., which added up to 21 risers and 20 treads. Though the rise and run weren't quite what the time-honored formulas sanctioned, in the end they worked just fine.

Now I knew how many treads to put into my sketch, but I still had to figure out their orientation and shape. This I did mostly by instinct. I put a tick mark every 12 in. along the path line in my sketch and used the marks as points on which to pivot the tread nosing. I simply tried to keep the taper of consecutive treads from changing too abruptly.

Once I was happy with the tread layout, I sketched in stringers as close to the ends of the treads as possible to avoid excessive cantilever. It took five pairs of stringers to skirt all the obstacles, and each of them was of a different length and slope (in any circular stair, inside stringers are steeper than outside stringers). The shortest stringers would carry three treads and the longest stringer would support seven treads.

**Placing the piers**—I knew where to place the piers to support the stringers, but I still had to determine their elevations. To do this, I calculated the elevations of all the treads above the shop slab and labeled them on my drawing. Then I drew a cross section of each tread and stringer that fell directly over a pier and figured as best I could the distance between the tops of the treads and the tops of the corresponding piers (the slopes of the stringers varied, so these measurements varied). By subtracting these numbers from the tread elevations, I obtained the heights of the piers.

With the drawing complete, it was time to return to the site and locate the piers, again using the deck framing as the point of reference. Then I shot the elevation of each one with the builder's level. I soon discovered that I was in for considerable digging. The required scraping simultaneously cleared the path for the lower two-thirds of the stairway and contributed fill to expand Ron's parking area.

Once I was convinced that the calculated pier elevations were correct, I poured the concrete piers and oriented wet post anchors in them to accept the stringers.

**Solving the stringers**—When the concrete set up, I began to work on the stringers. I worked from top to bottom so I could compensate for accumulated errors by adjusting the finished level of the concrete slab at the base of the stair (the accumulated error amounted to less than ⅛ in). The usual framing square and stair-gauge fixtures were useless for the layout of these stairs because of the different slopes of the stringers and because nearly every tread was a different shape. Instead, I laid out the stringers in place.

I started out by making plumb and level cuts on the ends of the upper pair of string-ers, figuring the cuts by taking the rise and

From *Fine Homebuilding* (February 1989) 51:46-49

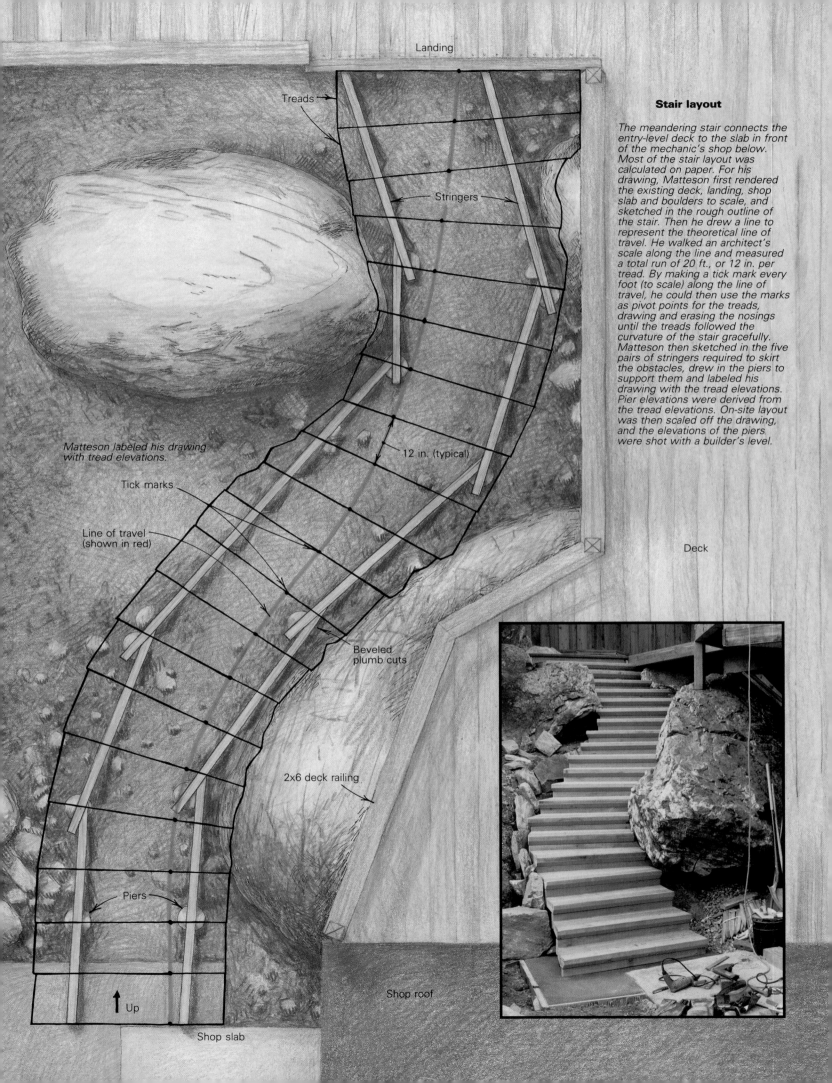

Landing

Treads

Stringers

### Stair layout

The meandering stair connects the entry-level deck to the slab in front of the mechanic's shop below. Most of the stair layout was calculated on paper. For his drawing, Matteson first rendered the existing deck, landing, shop slab and boulders to scale, and sketched in the rough outline of the stair. Then he drew a line to represent the theoretical line of travel. He walked an architect's scale along the line and measured a total run of 20 ft., or 12 in. per tread. By making a tick mark every foot (to scale) along the line of travel, he could then use the marks as pivot points for the treads, drawing and erasing the nosings until the treads followed the curvature of the stair gracefully. Matteson then sketched in the five pairs of stringers required to skirt the obstacles, drew in the piers to support them and labeled his drawing with the tread elevations. Pier elevations were derived from the tread elevations. On-site layout was then scaled off the drawing, and the elevations of the piers were shot with a builder's level.

Matteson labeled his drawing with tread elevations.

Tick marks

Line of travel (shown in red)

12 in. (typical)

Deck

Beveled plumb cuts

2x6 deck railing

Piers

Shop roof

↑ Up

Shop slab

run of each stringer off my drawing. After notching the bottoms slightly to clear the piers, I propped the stringers against a 4x6 header fixed to the landing and clamped them to the post anchors at the bottom. Next, I marked the inside of the stringers 10³⁄₆₄ in. down from the deck surface (the rise plus one tread thickness) and used my 2-ft. level to scribe a horizontal line through each mark. Going back to my drawing, I scaled off the run for the first tread at both stringer locations (the run differed on opposite ends of the tread) and marked their lengths on the horizontal lines. I used my rafter square to scribe a 6³⁄₆₄-in. long plumb line down from each mark, then drew another pair of horizontal lines, repeating the process until I reached the end of the stringer.

In order for the risers to lay snug against the stringers, I needed to bevel most of the plumb cuts. To figure these bevels, I simply extended my plumb lines up to the tops of the stringers, layed my 4-ft. level across the tops and aligned the edge of the level with a pair of plumb lines. Then I scribed a line across the tops of both stringers. That gave me the angles to which I adjusted my saw for each cut. I unclamped the stringers and cut along the lines with a worm-drive saw, finishing the cuts with a handsaw.

Next, I replaced the stringers, drilled pilot holes in them and screwed them to the deck posts and to the post anchors with #8 zinc-plated bugle-head drywall screws, greased with a dab of beeswax from a toilet gasket. I had special-ordered 2,000 of these screws for the deck from Mid-Valley Distributors (3886 E. Jensen Ave., Fresno, Calif. 93725). They worked so well that I decided to use them for the stairs, too.

I approached the rest of the stringers the same way as the first pair, except that the subsequent stringers each required a beveled plumb cut at the top end where they met the preceding stringers. To lay out the cuts for a pair of stringers, I notched the stringers to fit the piers and laid them in place, snugging the top ends up alongside the upper stringers. Then I used my 2-ft. level to draw plumb lines at the joint locations and used a saw protractor to figure out the proper bevels, transferring the bevels onto the stock. Once the stringers were marked, I took them down and put my worm-drive saw to work.

The required bevels for the top cuts were all greater than 45°, so I couldn't cut them in a single pass. Instead, I made the plumb cut with the saw set at 90°. Then I set the saw to 90° minus the bevel angle and made the finish cut by resting the base of the saw on the end grain exposed from the first cut. This was a little awkward, but it worked. Next time, maybe I'll clamp a block to the stock for added base support.

The varying width of the treads had two unavoidable consequences. Sometimes the treads dropped faster than the stringers, leaving less meat in the stringers than I liked. To stiffen up these skinny stringers, I thickened them with a pair of 2x6s fastened with con-

struction adhesive and screws. On some of the outside stringers, I encountered the opposite problem—the stringers descended faster than the treads, leaving a triangular void between the riser, tread and stringer (photo facing page, right). I addressed that problem by gluing and screwing vertical 2x6 blocks to the stringer to provide extra support for the front corner of the treads.

After cutting all the stringers to receive the treads, I screwed them to the post anchors and to each other, occasionally checking my progress down the slope with the builder's level.

**Treads and risers**—Starting again at the top of the stairs, I began to install the treads and risers. Each tread butts up against the riser above it and laps the riser below it.

Most of the treads were composed of three pieces: a 4x8 nosing piece, a 4x6 center piece (usually with an angle cut along the rear to fit against the riser) and a small wedge to fill out any remaining space along the back. I cut the nosing piece of the top tread to length first, ripped a 45° chamfer under its front edge so it wouldn't look too chunky, and smoothed out the chamfer with a power plane. Then I set the piece on the stringers so that the bottom of the chamfer would be flush with the front of the riser to be installed later below it, creating a 1¾-in. nosing.

With the nosing piece positioned on the stringer, the next step was to fit the middle piece (drawing facing page). To figure out the taper on the middle piece, I first held my tape measure perpendicular to the rear edge of the nosing piece and marked the point along that edge where it measured exactly 6 in. to the riser. That point would locate the beginning of the taper on the middle piece. Next, I measured the perpendicular distance between the nosing piece and the riser at its narrowest point. I transferred these measurements to the blank for the middle piece and cut the taper with the worm-drive saw. This method gave me a cutting line more easily, quickly and accurately than measuring the angles could have, and I repeated it for the wedge piece.

When the three pieces of the tread fit tightly in place, I reached underneath and scribed a pencil line along the outside edges of the stringers. Then I removed the pieces and eased their top edges with a block plane. I turned the pieces upside down and screwed a pair of 1½-in. by 1½-in. cleats to the two front tread pieces. The cleats were positioned along the pencil lines, and would later be screwed to the stringers horizontally to anchor the treads. I used redwood shims between the cleats and the tread pieces to account for the varying stock thickness.

Next I stood the partly assembled tread on its nose and further connected the two front pieces with three ½-in. by 6-in. lag bolts, counterbored into the back edge of the middle piece deep enough so that the entire threaded portion of the bolt would grip the

nosing piece. For all this hole-shooting, I rigged up a few drills: a Milwaukee Hole Hawg with an Irwin expansion bit (set at 1½ in.) for the counter boring and a ½-in. Milwaukee "Magnum" drill for the lag-bolt shanks and pilot holes, plus I had my dad's antique clunker for drilling all the pilot holes for the drywall screws. The Hole Hawg is frighteningly powerful; a few weeks earlier it snapped the ⁹⁄₁₆-in. shank of a self-feed auger in mid-hole with hardly a twitch. The Magnum drill is nice and sturdy. The only problem is that the handle is mostly switch, and it's hard to grip without turning it on. Once the holes were drilled for the lag bolts, I drove the bolts halfway home with an impact wrench. The last few turns went faster with a ratchet wrench, which also allowed me to feel when the bolts were snug.

I attached the wedge piece with drywall screws, which my cordless driver/drill sent burrowing into the redwood as far as I needed to counter sink them. That's another indispensable tool, especially if you have an extra battery charged up and ready to go.

Once the pieces were connected, I drove a few more screws into each through the bottom rail, flipped the tread over and used a round-over bit in my Sears Craftsman router to smooth the top front and side edges. During use, this router sucks bits up into the collet unless I really bear down with the collet wrench while changing bits—hard enough to bend the tooth on the shaft lock. To switch the bits after such punishment, the whole collet must be removed and the bit forced out with a nail set. Maybe I should just be happy the bit doesn't creep *down* during use.

To install the finished tread, I once again needed to use shims, this time between the tread and the stringers. When I was satisfied that the tread was at the proper height and solidly level, I glued the shims in place with construction adhesive and screwed the cleats into the stringers. I reached back under the tread and screwed the header above to it through the predrilled holes in the header.

When the tread was secured, I cut a riser to fit underneath it. I cut the riser to length, freshened up the face with the power plane and predrilled several holes along the bottom edge, which would later allow me to screw the riser to the back of the tread below it. Then I pried the riser tight against the tread with a flat bar and screwed it to the stringers with the drywall screws, low enough so the screws would be hidden by the tread below. I made and installed the rest of the treads and risers the same way (right photo, facing page).

Any time it was necessary to fit a riser or tread to a rock (left photo, facing page), I transferred the profile of the rock onto it with a contour gauge and gingerly trimmed the ends with a reciprocating saw and a sharp chisel. A contour gauge is made of two metal plates with about 150 tempered-steel wires sandwiched in between. When the gauge is pushed up against an object, the wires slide between the plates until they all touch the

Drawings: Christopher Clapp

object. The contour of the object can then be scribed off the gauge.

Making and installing each tread and riser pair took about two hours. After about a week, I reached the bottom stair and poured a small concrete pad at the base of the stairway.

**Finishing touches**—Once the stairs were finished, a coat of clear wood finish was brushed on (The Flood Co., P. O. Box 399, Hudson, Ohio 44236-0399). Ron built a nice, tight rock retaining wall against the edge of the stairs to keep dirt and critters from getting under them, and Liz landscaped very nicely around the deck and stairs. I got married and moved away to complete school. Those stairs were an enjoyable challenge and great practice for building an even nicer set of front steps for the Skeltons when I move back to the area.  □

*Thor Matteson is a civil-engineering student at California Polytechnic State University in San Luis Obispo, California. Photos by the author.*

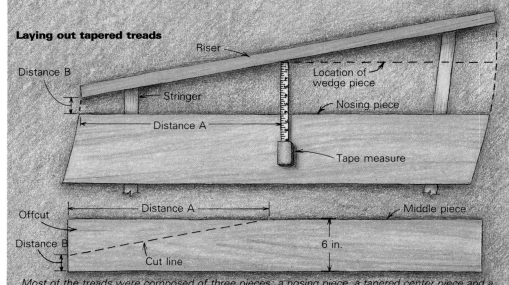

**Laying out tapered treads**

*Most of the treads were composed of three pieces: a nosing piece, a tapered center piece and a wedge piece to fill out the back. Once a nosing piece was cut and positioned on the stringer, the author held his tape measure perpendicular to the rear edge of the piece and marked the point where it measured exactly 6 in. to the riser. Then he measured from the end of the piece to the mark (distance A) and measured the gap between the nosing piece and the riser at the narrow end (distance B). He transferred these measurements to the blank for the middle piece, connected the points with a diagonal line and cut the taper with a worm-drive saw. The technique worked for the small wedge piece, too.*

The stair snakes around three boulders en route to the shop, requiring some careful fitting of treads and risers (photo above). The author used a contour gauge to transfer the profile of the stone to the tread and riser stock, then trimmed the stock with a reciprocating saw and a chisel. The stringers are made of 2x12 rough-sawn redwood (photo right), lapped at the joints and secured to wet post anchors and to each other with 3-in. zinc-plated drywall screws. The anchors are made to accommodate 4x stock, so redwood blocks were used to fill out the bottom anchors. Triangular voids in the left-hand stringer were remedied by screwing vertical 2x6 blocks to the stringers to lend extra support to the treads. The risers are screwed to the stringers with drywall screws, low enough so the screws are hidden behind the abutting treads, and they're also screwed to the backs of the treads. Chamfers on the treads prevent the stairs from looking too boxy. A concrete pad will later be poured at the bottom.

# Backyard Spiral

## A stair of redwood and steel built to endure the elements

by Richard Norgard

**B**efore I became a self-employed carpenter, I used to design and build scenery for the professional and educational theater. That's where I learned that spiral stairs date back to the Middle Ages and that, in those days, they typically coiled counterclockwise from top to bottom so that a right-handed defender with a sword could easily fend off a right-handed attacker coming up the stairs.

Working in the theater, I built several small, temporary spiral stairs, but I never constructed a permanent spiral stair. When Dan and Patsy Mote hired me to design and build a deck and a spiral stair onto the back of their house, I finally got the chance. The stair would not only be permanent, but it would also have to be built to beat northern California's weather.

**The overall plan**—The Motes' house, which is located in the Berkeley hills, is designed to accommodate large groups of people. The focal point for entertainment is the second-floor living room, which faces San Francisco Bay and overlooks a spacious backyard. The Motes wished to add a deck off the living room so they could entertain their guests both indoors and outdoors. But equally important, they wanted a new set of stairs to connect the deck with the backyard below.

After a series of discussions, we settled on a design that would go relatively easy on their pocketbooks and that would also accommodate future additions. The design called for a 200-sq.-ft. redwood deck to be built off the living room, with a spiral stair linking the deck to an existing deck in the backyard (photo above). Access from the house to the new deck would be via a pair of matching French doors that would replace an existing picture window in the living room. Instead of the usual piers, the deck would bear on a stepped concrete foundation. This would someday allow a sun room to be added beneath the deck. In the mean-

The spiral stair links the deck, which is attached to the main floor of the house, to the backyard. The stair consists of a column made of steel pipe, with steel brackets welded to the column, and redwood treads bolted to the brackets. Steel balusters support a glue-laminated redwood handrail. The stair rests on a 16-cu.-ft. chunk of concrete.

time, the foundation would contain a work area paved with brick embedded in sand.

Carpenter Tim White, assistant James Moore and I built the stepped foundation and deck with little trouble. We left the crux of the job, the spiral stair, for last.

**Designing the stair**—The Uniform Building Code only marginally covers spiral stairs. The code simply requires that treads measure 26 in. from the outside of the supporting column to the inside of the handrail, that the run is at

least 7½-in. at a point 12 in. from the narrow end of a tread and that the rise of the treads be no greater than 9½-in. (as opposed to the usual 8 in. for straight residential stairs). Regardless of the code, though, what's legal in one area may not be legal in another. Before I built my stair, I discussed in detail its materials and construction with the local building inspector and plan checker.

The final stair plan called for a column made of 3½ in. by ¼-in. thick steel pipe, with steel brackets welded to the column and redwood treads bolted to the brackets (middle drawing facing page). Square steel balusters would support a glue-laminated redwood handrail. Though my clients didn't plan to do any sword-fighting on their stair, a counterclockwise spiral fit best on the site.

The whole stair would weigh over 1,200 lb., enough to require a substantial footing to support it. We designed a massive reinforced-concrete slab that would both support the weight of the stair and anchor it to the hillside (bottom drawing facing page). The 16-cu.-ft. slab is shaped like an inverted truncated pyramid. It's reinforced with two horizontal triangles of ½-in. rebar, one above the other, with three vertical lengths of rebar tied to the points.

**Concrete and steel**—Because the stair landed on an existing deck, we had to remove the old decking before we could form and pour the slab. During the pour, we centered a ¾-in. thick by 12-in. by 12-in. steel plate on top of the slab and tied it to the concrete with four ½-in. anchor bolts. Once the concrete set, we positioned the steel column on the slab, plumbed the column and welded it to the plate. After the treads were installed, we drove a cylindrical piece of redwood into the top of the column and screwed it to the column with galvanized drywall screws (we call them "deck" screws). That allowed us to screw the deck handrail to the top of the column. The column would also be secured to a deck post below the joists with a pipe collar and a length of ½-in. threaded rod. A turnbuckle in the threaded rod would allow future adjustments. Once the column was welded in place, we reinstalled the decking on the existing lower-level deck.

The tread brackets were made by a local welding shop. Each bracket consists of a 3½-in. wide by 22¾-in. long piece of flat steel welded to a 7-in. length of steel tubing (top drawing right). The steel tubing was just the right size to allow us to slip the treads over the column later and weld them in place. Three 1-in. holes punched through each bracket with a punching machine would accommodate the bolts for attaching the wooden tread pieces to the brackets. To prevent rust, all steel in the stair, including the column, was finished with two coats of steel primer and three coats of heavy-duty exterior lacquer.

**Treading softly—**The tread pieces were cut out of select redwood 2x4s, except for the pie-shaped pieces on the outboard ends, which were cut out of 4x4s. We used redwood because it has excellent water-resistant and structural properties, it doesn't attract fungus or bugs and it's relatively inexpensive in northern California. By drawing a full-scale tread pattern first, we were able to determine the exact length and shape of each piece. We cut all the pieces on a 12-in. radial-arm saw and drilled the bolt holes using a drill press. We used a machinist's vise to position the pieces accurately.

Because this was an exterior stair, it was important that we provide gaps between the 2x4s of the treads for drainage. The galvanized washers commonly found in hardware stores are of an inconsistent thickness, so I used stainless-steel washers instead. Besides being of a uniform thickness (three of them created a gap of ⁵⁄₁₆-in.), they wouldn't rust or stain the wood. I found mine at Bowlin Equipment Co., 1107 10th St., Berkeley, Calif. 94710.

Once all the stair pieces were cut and drilled, we bolted them to the brackets with ½-in. galvanized carriage bolts, counterboring the bolt heads and nuts. Once the bolts were snug, we cut the ends flush using a reciprocating saw. Finally, we fixed a 2x4 cap to the end of each tread with galvanized drywall screws and sanded the tread tops with 36-grit sandpaper in a belt sander. This flattened the treads and created a slightly rough surface for better traction.

After sanding, the treads were ready for installation. As is often true with spiral stairs, the nosings (the overlaps of adjacent treads) on this stair taper. In this case, there's a 2-in. nosing at the column and no nosing at all at the outside edge. At first, the building department had a problem with this, but once they checked their catalogs and saw other stairs built this way, they acquiesced.

Using steel tubing as a collar worked out great. First we slipped all the treads over the

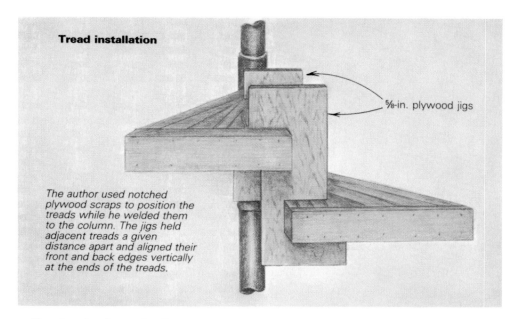

**Tread installation**

5⁄8-in. plywood jigs

*The author used notched plywood scraps to position the treads while he welded them to the column. The jigs held adjacent treads a given distance apart and aligned their front and back edges vertically at the ends of the treads.*

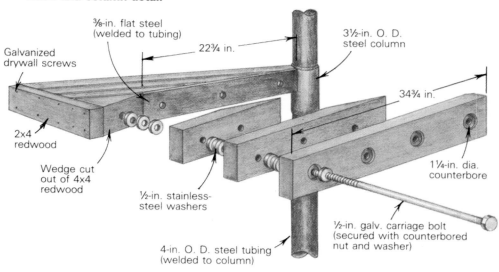

**Tread and column detail**

⅜-in. flat steel (welded to tubing)

3½-in. O. D. steel column

22¾ in.

Galvanized drywall screws

34¾ in.

2x4 redwood

1¼-in. dia. counterbore

Wedge cut out of 4x4 redwood

½-in. stainless-steel washers

½-in. galv. carriage bolt (secured with counterbored nut and washer)

4-in. O. D. steel tubing (welded to column)

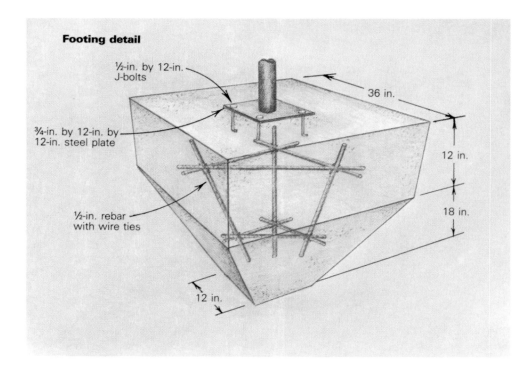

**Footing detail**

½-in. by 12-in. J-bolts

36 in.

¾-in. by 12-in. by 12-in. steel plate

12 in.

½-in. rebar with wire ties

18 in.

12 in.

The balusters (photo above) are made of square steel tubing with steel mending plates welded to the bottoms and a pair of angle-iron brackets welded to each top. The mending plates are lag-screwed to the backs of the treads and the handrail is attached to the top brackets with galvanized drywall screws. A lag screw anchors each baluster to an upper tread and eliminates vertical movement of the treads. The handrail is composed of six plies of redwood (photo left) which were glue-laminated on site. Aluminum connector bolts and cap nuts add a decorative touch and also help ensure against delamination. The photo shows the joint between the curved railing and a short horizontal return at the top of the stair. The rail is protected with three coats of spar varnish. *Photos by author.*

column and secured the pipe collar at the top of the column to the deck post. Then my associates lifted the top tread into position and held it there while I welded the collar to the column. The rest of the treads went up easily with the aid of a pair of jigs. The jigs were simply plywood scraps notched to hook over the nosing of a tread and to hold the adjacent tread a given distance below it (top drawing previous page). The jigs also served to align the front and back edges of adjacent treads vertically (at the end of the treads). Using a 2-ft. level to ensure that the treads were level, I welded the rest of the treads to the column. The final rise of each tread worked out to be 8⅝-in., well within code.

**Bending the rail**—The construction of the glue-laminated handrail both excited and concerned me (as glue-ups usually do). Because this railing was subject to both rain and sun, the materials, including the wood, glue and finish, had to be weather-resistant. For the sake of appearance, I also wanted the wood to be of a mixed color and free from knots. With the help of Jeff Hogan, owner of Ashby Lumber in Berkeley, I found what I

was looking for. Ashby Lumber carried pre-packaged "A" grade redwood boards which ranged in color from blond to reddish-brown. The boards were ⅜ in. thick by 3¼ in. wide and were surfaced on one side, and they came in random lengths (most of them measured between 8 ft. and 10 ft. long). I ripped the boards to 2⅜ in. wide on my table saw and thickness-planed them down to slightly more than ¼ in. thick. After I carefully arranged the boards according to color, they were ready for glue-up.

The finished handrails were to be 1⅝ in. wide (six plies) by 2¼ in. high. I used two-part resorcinol glue, which is waterproof, and staggered the butt joints between boards a minimum of 2 ft. For the glue-up forms, I temporarily attached to each tread a plywood gusset with a short length of 2x2 screwed to it. The gussets were attached to the treads with screws. Later, the screw holes would be covered by the balusters. Because the railing would have to be clamped every 4 in. or so, I collected every C-clamp I could find and then made the rest out of plywood scraps and carriage bolts with wing-nuts. The glue-up took three days to complete because I could man-

age only two plies per session without the glue setting up and without running into problems staggering the joints. Though the label on the can of glue recommended leaving the clamps on for 12 hours, I left them on for 24 hours.

Once the glue-up was finished, I removed the 27-ft. long handrail from the stair and planed the top and bottom smooth with a portable electric plane. I then eased the edges with my router chucked with a ¼-in. roundover bit.

**Building the balustrade**—The balusters were cut out of ⅛-in. thick by 1-in. square steel tubing (top photo left). I welded a ⅛-in. by 1-in. by 5-in. steel mending plate to the bottom of each baluster, extending the plate 3 in. below the end of the tubing. I cut the tops of the balusters at an angle to match the slope of the railing and welded a pair of pre-drilled 3-in. angle-iron brackets to the top of each baluster. I finished up by bending the brackets to the proper angle and rounding their corners with a grinder.

To install the balusters, I rested them on the treads and screwed the mending plates to the back edges of the treads with 5⁄16-in. by 3½-in. lag screws fitted with lock washers. Each baluster touched the nosing of the tread above and was screwed to it with the same lag screws and lock washers. This totally eliminated vertical movement between treads.

After bolting the balusters to the treads, I positioned the handrail on top of the balusters and scribed the shape of the angle-irons on the underside of the railing. Then I removed the railing and by eye routed out mortises for the brackets using a router and a ½-in. mortising bit; I wanted the brackets to be flush with the bottom of the railing. Because the corners of the angle iron were rounded, there was no need to chisel out the corners of the mortises. Once the routing was complete, I repositioned the railing and secured it to the balusters with 1⅝-in. long galvanized drywall screws. Up top, the transition to a short horizontal railing section gave us the chance to display some tidy joinery (bottom photo left).

As a decorative touch and an extra precaution against delamination, I drilled ¼-in. holes through the railing and installed aluminum joint-connector bolts and cap nuts on 6-in. centers. These bolts can be adjusted from 1¼ in. to 2 in. in length and come in brass, aluminum or stainless steel. The bolts added a distinctive nautical appearance to the stair. I got mine at a local boat shop.

Finish-sanding of the railing went quickly. As the final step, I brushed on three coats of high-gloss spar varnish. This not only assured retention of the color in the wood, but also eliminated any possibility of splintering. □

*Richard Norgard is a free-lance carpenter in Fort Bragg, California.*

# Single-Ply Roofing
## Tough, versatile membranes tackle the toughest roof problems

by Harwood Loomis

Since man first sought shelter from weather and carnivorous neighbors, he has preferred to have a roof over his head. Historically, the form of the roof has ranged from animal skins and thatched tree fronds to modern synthetic materials. The material most used in the U. S. for residential roofs these days, however, is probably the ubiquitous asphalt shingle. Other roofing materials include slate, clay, terra-cotta and concrete tiles, metal sheets, and wood shingles and shakes. All these materials have one thing in common: they are generally suitable only for use on roofs having a slope of 4-in-12 or more. At slopes below that, the designer and contractor have, until recently, had only one serious choice: the built-up roof, a multi-layer sandwich of roofing felt and either asphalt or coal-tar pitch (rolled roofing should only be considered for outbuildings and utility sheds).

Today, however, the conventional built-up roof is being eclipsed by a relatively new generation of sheet products called "single-ply membranes." Though associated mostly with huge commercial roof projects, single-ply membranes are now finding considerable use in residential construction.

The term single-ply membrane includes a vast number of different systems and materials, some of which are actually made up of more than one ply (photo next page). They're called single-ply because they are generally *installed* as one ply, rather than layered on like traditional roofing materials.

A single-ply roof is typically sold as a system of proprietary fasteners and adhesives which, along with the membrane itself, are applied by contractors licensed by the membrane manufacturer. It's possible for a builder to install a small job with materials purchased from a roofing contractor, but the membrane manufacturer typically won't extend any warranty on the materials in such cases (one exception is Resource Conservation Technology—see sidebar, p. 63).

**Why single-ply—**New single-ply membranes seem to come on the market daily, and there are easily more than 300 products to choose from. One of the attractive aspects of single-ply roofing in general is that it does away with the need for the smelly tanker full of molten asphalt that's needed for the installation of built-up roofing. And by getting rid of the ket-

Single-ply membranes are particularly useful in sealing flat roofs over living spaces. In the photo above, an EPDM membrane covers a sunroom addition to a large house.

tles, builders also get rid of the open flames used to heat them, making a safer job site. But there are other advantages as well.

Designers and builders of custom homes no longer need to have their imaginations limited by the roofing technology of decades past. Many single-ply products are elastic and allow roofs to take on complex, compound curves or to be nearly flat, yet still be watertight and durable. Some of the systems are well suited to double duty as roof terrace walking surfaces; others can serve as a waterproofing substrate beneath wood decks. Single-ply membranes are also effective when used to waterproof pocket decks and additions with flat roofs (photo above). Very probably, one of these products will be suitable for any roof configuration where shingles, slates or tiles can't do the job.

**An alternative to built-up—**To appreciate the popularity of single-ply roofing these days, it helps to know something about the evolution and use of built-up membranes, the traditional product for covering low-slope roofs. This is what some people refer to as a tar-and-

gravel roof, although it may not be made up of tar and it may or may not be surfaced with gravel. Typically it was composed of layers of roofing felt alternated with mopped-on layers of either melted asphalt or coal-tar pitch. This "old faithful" in its several variations has served builders well, but it has some drawbacks.

One problem is that the raw materials available today are not as good as they were in the past. The most common component of a bituminous roof is asphalt, which comes from various grades of crude oil. With prices rising as fast as they have lately, coupled with diminishing reserves, there is more competition to refine as much "product" as possible from the raw crude. Roofing asphalt comes out low on the totem pole in comparison to products such as gasoline, diesel fuel and heating oil. Many roofing experts believe that by the time all the other products have been drawn off, there isn't much left of what used to make roofing asphalt a satisfactory weatherproofing medium.

The best of the old tar-and-gravel roofs were made of coal-tar pitch. These roofs were excellent; I've inspected many that lasted 40 or

50 years before beginning to leak. Unlike asphalt, coal tar becomes self-healing in warm weather—the warmth softens the coal tar enough to seal any cracks—so the roofs tended to repair themselves every summer. Unfortunately, the fumes from coal-tar pitch heated during installation have proven to be carcinogenic. This led the government to place strict limits on amounts and concentrations of fumes that could be given off during the installation of a coal-tar roof. That, in turn, resulted in changes to the formulation of the pitch itself, which diminished its durability. Coal tar isn't plentiful today, and most roofers and roofing consultants agree that what *is* available is not as good as the product of 40 years ago.

Another problem with a bituminous roof—perhaps "challenge" is a better word—is that residential designers today are creating roof shapes that are more irregular and complicated than they used to be, and they're coming up with all sorts of other low-slope roofs, including some with compound curves. The traditional bituminous roof just isn't up to covering shapes like these, for good reason. The strength of a bituminous roof is imparted by the layers of standard 15-lb. roofing felt that are rolled over the roof between moppings of melted bitumen (asphalt or coal-tar pitch.) The felts come in rolls 36 in. wide and roll out very nicely on a regular, flat roof deck. But they do not readily lend themselves to rolling out on curved or compound-curved surfaces.

Given the increasing problems with built-up roofing, it's no surprise that single-ply products are taking over the market for low-slope roofing. Here's a look at the major types of single-ply roofing membranes that have applications for the residential market.

**Rubber membranes—**One of the most popular materials for single-ply roofing is synthetic rubber. There are a number of synthetic-rubber products on the market, two of which seem to have established themselves as reliable, high-quality materials for roofing membranes: EPDM (ethylene propylene diene monomer) and CSPE (chlorosulphonated polyethylene). Hypalon, DuPont's tradename for CSPE, has become virtually synonymous with CSPE roofing. Both are marketed by major manufacturers and are available with a choice of several different installation systems to address various roof configurations.

EPDM comes in a black sheet that looks and feels like a tire inner tube (photo above). It's usually 45 mils thick (a mil is $\frac{1}{1000}$ in.) and weighs about .28 lb./sf. Sheets range from 10 ft. to 100 ft. wide. A sheet of EPDM is usually not reinforced with fibers, and that allows it to stretch somewhat to accommodate irregular roof shapes. EPDM is reasonably resistant to many chemicals but is susceptible to attack

Though there are many kinds of single-ply membranes, the ones most commonly used for residential applications are shown above. From front to back: MBR (this one is styrene butadiene styrene), PVC, EPDM, reinforced PVC, EPDM, and CSPE.

by organic fats (such as cooking oils), some solvents (aromatic hydrocarbons like gasoline and oil), and fire. A special formulation of the product, sometimes called FR sheet, is available for projects in which the membrane surface will be uncovered and where a fire-resistant roof is required.

EPDM membranes aren't available in colors, but they can be coated with liquid Hypalon rubber, which comes in a range of colors. Some manufacturers offer a white EPDM sheet, but they won't warrant it for as long as they would an otherwise identical black membrane. That's because the material that makes EPDM black—carbon black—also contributes to its durability; white membranes lack this element.

**Installing EPDM—**An EPDM roof may be installed in several ways. In a *ballasted* application, a large sheet of the membrane is laid over the roof deck and smoothed into place. A heavy fabric is placed over the rubber to protect it, and the assembly is then covered with ballast—washed, round stones or cast-concrete pavers—that keeps the membrane from lifting off the roof in high winds and protects it from fire. The ballast generally weighs 10 lb. to 15 lb. per square foot. The perimeter of the roof and any penetrations, such as chimneys or plumbing vents, are sealed with special adhesives and flashing material (an uncured and elastic form of the EPDM rubber itself). Seams in an EPDM roof are sealed with adhesive.

The advantages of this system are fast, simple installation and relatively low cost. The disadvantage is that the house must be strong enough to support the weight of the ballast material in addition to whatever loads have to be anticipated for wind, rain, snow and people. The ballast also makes it difficult to find and repair leaks, and it is often difficult to obtain round, unbroken stones in the size needed for ballast.

A second method for installing an EPDM membrane is *mechanical fastening*. There are a number of different systems for mechanical attachment, athough most are proprietary. The EPDM membrane is rolled out over the roof deck and fastened to the structure at regular intervals with either batten strips or some sort

of metal or plastic fastener plate that is screwed to the roof deck.

The advantages of mechanical fastening systems are their relatively fast installation, moderate cost, light weight and the ease of finding leaks. A disadvantage is that wind action on the loose membrane may cause the fasteners to loosen and "back out" of the roof deck, causing localized fatigue of the membrane at the fasteners.

The third major method of installing EPDM membrane is called *fully adhered*. Rigid insulation board is first installed and firmly attached to the roof deck, usually by screws. The membrane is then rolled out and fastened to the insulation board with proprietary contact adhesives specially formulated for the purpose. These roofs are susceptible to wrinkles during installation, so a heavier membrane (60 mil) is used.

Like other installation systems, the fully adhered approach isn't perfect. Advantages include light weight and the ease of locating and repairing leaks. On the downside, the fully adhered system is the most expensive of the three methods and takes the longest to install. It leaves the membrane exposed to fire and physical damage and calls for more field seams than with other systems. This is because of the difficulty in working with wide sheets coated with adhesive—it's pretty tough to keep from getting wrapped up in your work. So manufacturers of fully adhered systems limit the sheet to a 10-ft. width. And if the roof will be a visible design element, high quality workmanship becomes very important; it's easy to wrinkle a fully adhered system while installing it.

The fourth method of installing an EPDM roof is called a *protected-membrane* system. Also called an IRMA system (Inverted Roof Membrane Assembly), it's basically a variation of the ballasted roof and is sometimes called an "upside-down" roof. The membrane goes over the roof deck, then Styrofoam insulation at least 1 in. thick goes over that. The Styrofoam is held in place with ballast; channels molded into the bottom of each Styrofoam sheet allow water to drain freely from the roof. The big advantage of this system is that it protects the membrane from thermal shock (stresses induced by daily and seasonal temperature swings). Thermal shock will speed the eventual deterioration of any roof membrane.

**The CSPE membrane—**DuPont's Hypalon is the best-known member of the CSPE family. Like EPDM, it is a synthetic rubber membrane, but Hypalon is reinforced with a polyester scrim, an open grid of fibers. The reinforcement increases the tensile strength of the membrane, but reduces its elasticity as well, making it less desirable than an unreinforced product for use on roofs with compound curves.

The standard white color of Hypalon (other colors are available) reduces rooftop heat gain, but white is also difficult to keep clean in dirty, industrial sections of the country. Hypalon should not be selected purely for appearance reasons unless the homeowner is willing to expend some effort to keeping the roof surface clean (periodic scrubbing with household detergent should do it).

Hypalon roofs can be installed with any of the methods used to install an EPDM roof (ballasted, mechanically fastened, fully adhered and protected membrane). Rather than being bonded with adhesives, however, Hypalon seams are welded together. A hot air gun is used to soften the rubber, and the seams are rolled together to fuse the joint.

Hypalon is very resistant to a wide variety of chemicals, but it's susceptible to attack from oils and gasoline. Hypalon costs somewhat more than EPDM and requires more field seams (the widest Hypalon sheets are generally 5 ft. wide, while the *narrowest* sheets of EPDM are usually 10 ft. wide). It weighs about .29 lb./sf.

**Other rubber membranes—**There are a number of other synthetic rubber materials used as roofing membranes, although none seem to have achieved the widespread acceptance of EPDM and CSPE. Nonetheless, each has advantages under certain conditions. Among the other products are: CPE (chlorinated polyethylene), PIB (polyisobutylene), and EIP (ethylene interpolymer alloy).

A thermoplastic, CPE comes from the same base as CSPE, but it doesn't contain sulphur. It is extremely resistant to most chemicals. Field seams and repairs are heat-welded, and repairs don't require liquid reactivators to soften the original membrane. A CPE job requires that metal flashings be precoated with liquid CPE. Only one company sells this product right now, and they don't issue warranties on residential applications.

A PIB membrane comes as a 100-mil sheet

Most single-ply membranes can be applied without the need for heat. One type of modified bitumen membrane, however, employs heat to liquefy and adhere the underside of the membrane to the substrate.

with a fleece backing that can be set either into hot asphalt or a cold adhesive. It's compatible with asphalt, and relatively simple to install. The sheets have 2-in. wide self-adhering tape built into the edges; to seal a seam, a release paper is pulled away, the edges of the sheets are overlapped and the seam is pressed with a metal roller.

Like Hypalon, an EIP membrane is made from DuPont polymers. It's a thermoplastic consisting of large polymer molecules that, according to the manufacturer, lock in the plasticizer molecules to reduce plasticizer migration (more on this later). Field seams are made by heat welding; repairs can be made by heat welding without requiring the use of liquid reactivators. The system is available only for mechanical attachment.

**Plastic membranes—**With continued advances in the composition of roofing membranes, it's getting tough to differentiate between "rubber" and "plastic" membrane materials. While EPDM and CSPE are almost universally accepted as rubber, some of the other membrane materials listed above can be

argued to be plastics rather than synthetic rubbers. But the distinction between rubber and plastic, as far as I am concerned, is that a plastic membrane requires chemical additives called plasticizers to remain flexible. These plasticizers gradually outgas over time, and if plasticizer migration is excessive, the membrane may become brittle and crack.

One membrane that absolutely belongs in the plastic category is PVC (polyvinyl chloride). The earliest PVC roofs in this country were characterized by a loss of plasticizers and a resultant brittleness and shrinkage. They sometimes failed prematurely either because they cracked, or because they shrank and pulled away from the perimeter of the roof. Manufacturers of PVC membranes have devoted considerable time and effort to optimizing the amount and composition of the plasticizers they use, and claim to have corrected any problems. While this may be true, one major supplier of PVC membranes limits the warranty to 10 years and the other limits it to 15. By contrast, most EPDM suppliers will back warranties of up to 20 years for commercial installations.

A PVC roof can be installed using the same techniques as those used with EPDM roofing (although one major supplier does not include a fully adhered system in its line). Seams are either fused with solvents or welded with hot air. Like EPDM and CSPE, PVC roofs are lightweight (except for the ballasted systems) and are easy to install. Cost is generally in the same range as EPDM.

Another problem with PVC membranes, however, is that they are less resistant to chemicals than EPDM or CSPE. In addition to being susceptible to attack from oils and animal fats, PVC can also be damaged if exposed to bitumens used in conventional roofing. This means that if PVC is used to reroof an area already covered with asphalt shingles or con-

## Sources for single-ply membranes

For an excellent guide to single-ply membranes, contact the National Roofing Contractors Association (10255 W. Higgins Rd., Suite 600, Rosemont, Il. 60018; 708-299-9070). Their "Roofing & Waterproofing Manual" contains just about everything you'll need to know about roofing, including single-ply membranes. The manual costs $118 (non-members).

The Single-Ply Roofing Institute (104 Wilmont Rd., Suite 201, Deerfield, Il. 60015-5195; 708-940-8800) has a comprehensive list of various manufacturers, categorized by membrane type. There are too many

manufacturers to list in this article, but here's a selection of manufacturers whose products are generally available. —H. L.

### EPDM membranes
Firestone Building Products Co., 525 Congressional Blvd., Carmel, Ind. 46032; 800-428-4442

Resource Conservation Technology, Inc., 2633 N. Calvert St., Baltimore, Md. 21218; 301-366-1146

The Goodyear Tire & Rubber Co., Roofing Systems, 1144 E. Market St., Akron, Ohio 44316; 800-992-7663

Manville Corporation, Roofing Systems Div., P. O. Box 5108, Denver, Colo. 80217; 800-654-3103

### CSPE membranes
JPS Elastomerics Corporation, P. O. Box 658, Northampton, Mass. 01061-0658; 413-586-8750

### MBR membranes
U. S. Intec, Inc., P.O. Box 2845, Pt. Arthur, Tex. 77643; 800-624-6832

Owens-Corning Fiberglas Corp., Commercial Roofing, Fiberglas Tower, Toledo, Ohio 43659; 419-248-8000

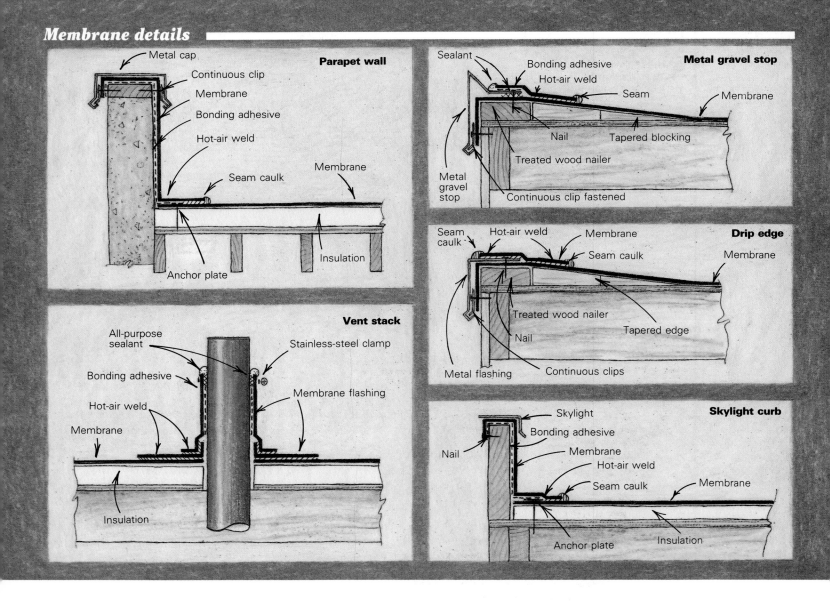

**Parapet wall**
- Metal cap
- Continuous clip
- Membrane
- Bonding adhesive
- Hot-air weld
- Seam caulk
- Membrane
- Insulation
- Anchor plate

**Metal gravel stop**
- Sealant
- Bonding adhesive
- Hot-air weld
- Seam
- Membrane
- Nail
- Tapered blocking
- Treated wood nailer
- Metal gravel stop
- Continuous clip fastened

**Drip edge**
- Seam caulk
- Hot-air weld
- Membrane
- Seam caulk
- Membrane
- Treated wood nailer
- Tapered edge
- Nail
- Metal flashing
- Continuous clips

**Vent stack**
- All-purpose sealant
- Stainless-steel clamp
- Bonding adhesive
- Membrane flashing
- Hot-air weld
- Membrane
- Insulation

**Skylight curb**
- Skylight
- Bonding adhesive
- Nail
- Membrane
- Hot-air weld
- Seam caulk
- Membrane
- Anchor plate
- Insulation

ventional built-up roofing, extreme care must be taken to ensure that no asphaltic products or debris are tracked onto the PVC membrane. A separator sheet should be used beneath the PVC membrane to ensure that it doesn't come in contact with the original membrane. Perimeter flashings must be made using special metal accessories precoated with PVC to which the roof sheet can be fused.

While the material comes in white, grey, and some colors, it tends to attract atmospheric dirt almost magnetically, so it can be difficult to keep clean where appearance is important. It might seem that the appearance of a roof membrane on a flat roof is less of a factor than the appearance of a gable-roof covering. But when an adjacent portion of the building overlooks the membrane, the view below shouldn't detract from the view beyond.

**Modified bitumen membranes** — As the quality of asphalt bitumen for roofing deteriorated, synthetic rubber and plastic membranes captured ever larger shares of the roofing market. The asphalt industry decided to fight back by researching ways to improve the performance of the asphalt. What they did was to modify raw asphalt with chemicals, resulting in a rubberized asphaltic material generally called "modified bitumen." The product is combined either with fiberglass or with non-woven polyester reinforcement to form a 160-mil roll product that's installed as a single layer, rather than being hot mopped in several succeeding layers as with conventional built-up roofing. With the reinforcement, the product is called MBR (modified bitumen, reinforced).

MBR membranes are extremely tough and puncture-resistant, so they're well suited to areas where there's the possibility of falling tree branches and other physical projectiles. At 1 lb./sf, they're about three times the weight of CSPE and EPDM, but are considerably lighter than a built-up roof (a 4-layer smooth surface built-up roof weighs about 1.5 lb./sf without gravel and 5 lb. to 6 lb./sf with gravel). Like other types of single-ply membranes, MBR membranes may be applied to roofs of any slope, from dead flat to absolutely vertical. An MBR membrane generally costs slightly less than rubber or PVC membranes.

Field seams in MBR roofs are made with the same techniques used to install the membrane itself: heat welding, hot mopping or cold adhesives. The basic color of MBR sheets is black, but some manufacturers offer membranes with a colored granular surface. Loose granules in a matching color can be sprinkled onto the field seams to achieve a uniform color over the entire roof surface.

Because the asphalt is modified, MBR sheets are more flexible than sheets of conventional roll roofing and can be formed around corners without tearing, particularly when they're softened with a torch. But MBR sheets do have limits, and can be difficult to form into the complex shapes that are needed to flash some penetrations. Perimeter and penetration flashings are made using the same membrane used for the roof. Like conventional built-up roofs, MBRs are susceptible to concentrated acids and to hydrocarbons.

Unfortunately, some MBR roofing requires the use of installation techniques reminiscent of the old hot-kettle methods used for built-up roofs, methods that other single-ply membranes have done away with. Several different types of modified bitumens are available today, but two types predominate: APP (axactic polypropylene) and SBS (styrene butadiene styrene). The difference between them is largely the method of installation.

APP modified bitumen membranes are "heat applied." A roll of membrane material is posi-

Drawings: Christopher Clapp

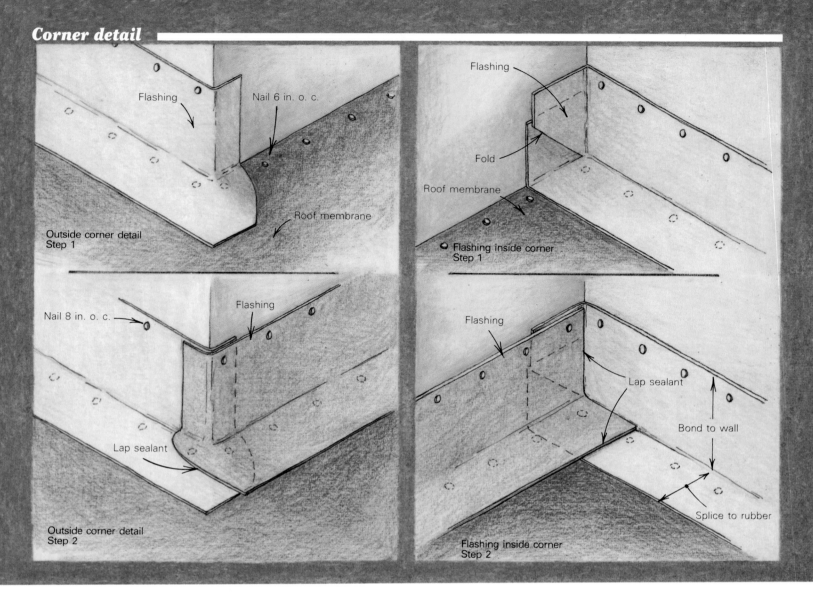

Outside corner detail
Step 1

Flashing

Nail 6 in. o. c.

Roof membrane

Flashing

Fold

Roof membrane

Flashing inside corner
Step 1

Nail 8 in. o. c.

Flashing

Lap sealant

Outside corner detail
Step 2

Flashing

Lap sealant

Bond to wall

Splice to rubber

Flashing inside corner
Step 2

tioned so that it will unroll into its final position on the roof. The underside of the roll is then heated with an electric heating element or a gas torch until the material softens and begins to flow. Then the membrane is rolled out, and the melted underside becomes the adhesive (photo, p. 63). For small roofs the torch may be hand-held, but for large roofs there are special carts to hold the roll and a built-in row of torch nozzles so that a uniform flame is applied across the entire width of the roll. The equipment and the process are proven, and they're safe in the hands of skilled and careful installers. But the process requires high temperatures, so torch-applied MBR roofing is usually not recommended over wood decks or combustible insulation materials.

An SBS membrane may be applied using a flood coat of conventional hot asphalt as an adhesive, or by using a proprietary cold adhesive. The hot-mopped installation brings back the asphalt kettle and open flame associated with built-up roofing, and so may not be attractive to people who were glad to bid adieu to such equipment. In addition to being smelly and dangerous, hot asphalt is just plain messy. It's tough to use without spilling some of the black goo where it is least wanted.

Torch-applied APP, even though subject to some possibility of fire as a result of the torches, is neater than hot-mopped SBS. Adhesive applied SBS is equally neat, and does not require any open flames on the job site.

**Application is for the pros**—While almost anyone can buy a plug of asphalt and some rolls of roofing felt in order to do a built-up roof, most of the single-ply products are available only to franchised applicators. While this means that a builder generally can't buy the material and do it himself (an approach I don't recommend, because it precludes any warranty from the manufacturer), it also means that when you find a licensed applicator, the full resources of the manufacturer are also available for technical advice.

If you insist on working with the material yourself, and some builders do, you won't find it at the local lumberyard. Instead, check with a roofing wholesaler, or go to a franchised applicator and purchase surplus sheeting. Remember, though, that with CSPE and EPDM membranes, you'll also need the compatible adhesives and accesories. The drawings above show samples of membrane details commonly encountered in residential work.

Patching a hole in an EPDM membrane is just about like patching an inner tube. After you find the leak, cut a patch, clean the surrounding area and fasten it down with adhesives. CSPE and PVC membranes have to be chemically treated before a patch can be applied, however, because the membrane itself has to be softened up in order for the patch to be heat-fused on. Any serious repairs of single-ply membranes are best left to those who put the system in.

One design note. Any kind of single-ply membrane makes an excellent vapor barrier, so make sure there's plenty of air circulation between any insulation and the membrane to prevent moisture from being trapped in the roof assembly. Alternately, you can keep moisture from the roof cavities by providing a vapor barrier beneath the interior ceiling surfaces. Protected membranes, however, are one exception. Because this puts the membrane on the warm side of the insulation, there's no need for a separate vapor barrier. □

*Harwood Loomis is a consulting architect specializing in building envelopes, roofing and waterproofing. Photos by the author except where noted.*

# Venting the Roof

## With today's construction methods, the minimum code requirements may not be enough

by Kevin Ireton

**O**n the subject of roof ventilation, the Canadians are fond of quoting Gus Handegord, one of their premiere building scientists. Handegord says, "If you go into an attic in the winter, find frost on the underside of the sheathing and there are no vents, you should put some in. But on the other hand, if you go into an attic in the winter, see frost on the the underside of sheathing and there are vents, close them up."

Although his advice sounds contradictory, Handegord simply recognizes that attic ventilation is a relative issue. How, where, when and even whether you should vent your roof depends on the type of construction, level of insulation, climate, site and probably a dozen other variables. The truth is, very little research has been done on roof ventilation, and the experts don't always agree on what's best.

In this article the term "roof ventilation" refers to a vented air space over the insulation in an attic or a cathedral ceiling. Roof ventilation should not be confused with house ventilation. In a tightly built house mechanical ventilation is needed to expel stale air and to bring in fresh air for breathing and combustion (for more on mechanical ventilation see *FHB* #34 pp. 30-34).

**Codes and warranties**—All the building codes require roof ventilation. Most of them enforce 1:150 and 1:300 ratios. According to these ratios, you need a *minimum* of 1 sq. ft. of net free vent area (NFVA) for every 150 sq. ft. of ceiling area below the roof. (The NFVA is the total area of the vent opening and takes into account the area blocked by any screening or louvers.) If you use a vapor barrier on the ceiling, however, or if you divide the ventilation evenly between high and low vents, then the minimum is 1 sq. ft. of NFVA for every 300 sq. ft. of ceiling. The vapor barrier reduces the amount of moist air infiltrating the attic, hence reducing the need for ventilation. Dividing the ventilation between high and low vents increases its efficiency, which also reduces the amount of ventilation needed.

Manufacturers of asphalt and fiberglass shingles also agree that ventilation is a good idea, at least for the longevity of their products. Ventilation under the roof deck (to reduce heat buildup) is a condition of shingle warranties. Some manufacturers (Georgia-Pacific, for instance) even go so far as to stipulate the 1:150 ratio.

**A little history**—Prior to the 1930s, people didn't vent their roofs. They didn't have to. Roofed with wooden shingles over open sheathing, and with little or no insulation in the attic, houses weren't tight enough to trap heat or moisture. But the next 60 years saw the advent of building papers, asphalt shingles, plywood, insulation and vapor barriers, which had the same effect as putting a lid on a pot of boiling potatoes. As a result, mois-

As a result of warm air leaking from the house into an unvented attic in Manitoba, Canada, severe frost accumulated on the underside of this roof sheathing. When the frost melted, it literally rained in the attic.

*The ice dam cometh.* As heat from the house leaks through the ceiling and up to the roof, it melts the bottom layer of snow. The water flows down to the cold eave, refreezes and forms a dam. Subsequent melting forces water and ice to back up under the shingles.

ture—from breathing, cooking, bathing, plants—began to accumulate inside houses. Households weren't generating more moisture, but tighter houses were retaining more of it.

By the 1940s problems with mold, mildew and peeling paint were common enough that Frank Rowley at the University of Minnesota was investigating the use of ventilation to remove moisture from attics. It was his research that eventually led to the adoption by the FHA and subsequently by the building codes of the 1:150 and 1:300 ratios. These were the levels of roof ventilation deemed necessary for the leaky, poorly insulated houses typical of the 1940s. And these ratios served fairly well for the next 30 years.

Since the energy crisis in the 1970s, however, builders and remodelers have been tightening walls and ceilings and adding still more insulation. This means that attics are colder and that any house air escaping into them is likely to have a much higher moisture content than ever before—the ideal conditions for condensation. Curiously, the building codes have not been revised to reflect these changes.

**Venting for moisture reduction**—Moisture buildup in a roof can lead to peeling paint, wet insulation, the growth of mold and mildew, and ultimately to failure of the roof structure. Architects, builders, building scientists and vent manufacturers may argue about ventilation, but they all agree that keeping moisture out of the roof in the first place is the best way to eliminate problems. And by now most of them agree that air-transported moisture causes more problems in roofs than does moisture from any other source (such as vapor diffusion).

If you have a tight house with lots of insulation, you must seal up air leaks from the house into the roof. This doesn't mean you have to use a polyethylene vapor barrier. Carefully installed, drywall does just fine at stopping air leaks (for more on this see *FHB* #37, pp. 62-65). But chimneys, partition walls, holes drilled for electrical and plumbing runs, fiberglass tubs with integral surrounds, recessed can lights, electrical boxes and attic hatches all allow moist house air to flow into the attic. And holes cut for plumbing vents are probably the worst culprits. They should be sealed with caulking and a gasket made from a piece of rubber (EPDM) membrane.

Top photo by Harold Orr, courtesy of the Institute for Research Construction, National Research Council of Canada

If warm, moist, house air does leak into the attic, ventilation air will carry it out before it has a chance to condense on the cold underside of the sheathing (top photo, facing page). But as air gets colder, its ability to absorb moisture is reduced. So as you increase the insulation levels in the ceiling, you make the attic air colder and decrease its ability to carry off moisture.

What should you do? Joe Lstiburek, a Canadian building engineer, says that if you live in a superinsulated house (with ceiling insulation levels of R-45 or more) in a very cold climate, you should close your roof vents during the winter. Clarke Wolfert, a professional engineer and technical director for Air Vent, Inc., simply says you need to increase the ventilation rate in the attic and move more air through it.

**Venting to prevent ice dams**—Ice dams form when snow on the roof melts, flows down and freezes again as it hits the relatively colder eaves. The accumulation of ice actually forms a dam, and subsequent melting forces water and ice underneath the shingles behind the ice dam (bottom photo, facing page). If a house is well insulated, losing little or no heat through the ceiling, then the snow on the roof won't melt…at least not right away.

Problems arise as snow accumulates on a roof; if it's relatively dry, snow is a good insulator—about R-1 per in. Accumulating snow insulates the roof surface, which means that even the slightest heat loss (i.e., thermal conduction) can raise the temperature of the roof surface above freezing. This melts the snow sitting directly on the roof, which then flows to the eaves where it freezes. Proper ventilation creates a thermal break and circulates cold air under the roof, keeping the surface of the roof colder and preventing the snow from melting.

**Venting to reduce heat buildup**—The third basic reason for venting roofs is to reduce the buildup of heat in hot weather. Especially when used in conjunction with a radiant-barrier system in southern climates, roof ventilation significantly lowers the cooling bill and raises the comfort level of houses. But studies done at the Florida Solar Energy Center (FSEC, 300 State Rd. 401, Cape Canaveral, Fla. 32920) have shown that shingle and sheathing temperatures are far more dependent on shingle color than they are on ventilation. So if you're really worried about heat buildup, use light-colored shingles in addition to ventilation.

Can moisture, in the form of hot, humid air, actually enter the attic through the vents? If the house is air-conditioned, will moisture then condense on the back of the ceiling drywall? According to Philip Fairey, principal research scientist at the FSEC, the answer is yes, if the thermostat is set low enough. With the air-conditioning set at 77° or above, however, the chances of condensation problems are negligible.

Lstiburek says, just in case, use 1-in. thick

Drawings: Michael Mandarano

From *Fine Homebuilding* (June 1990) 61:76-80

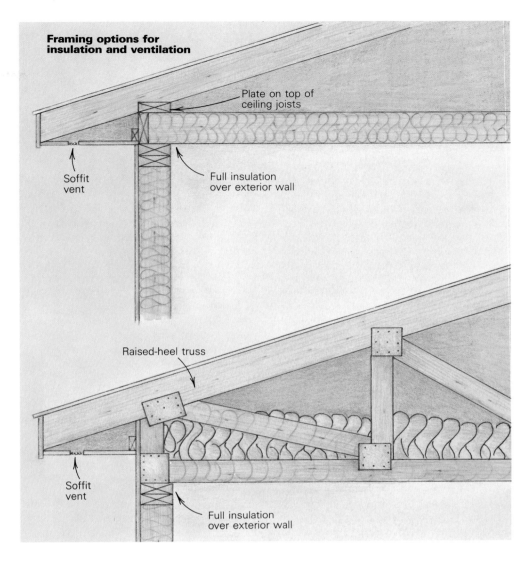

**Framing options for insulation and ventilation**

Plate on top of ceiling joists

Soffit vent

Full insulation over exterior wall

Raised-heel truss

Soffit vent

Full insulation over exterior wall

vapor-impermeable insulating sheathing (foil-faced rigid foam) behind the ceiling drywall, if you live in a cooling climate. Such sheathing would stay relatively warm on the attic side and reduce the chances of condensation.

**Warm roofs, no vents**—Bob Corbett, the senior technical specialist with the National Center for Appropriate Technology (P. O. Box 3838, Butte, Mt. 59702-3838) says that seven years ago he built a house in Montana with a flat, unvented roof and has had no moisture problems. John Abrams, a designer and builder on Martha's Vineyard has been building houses with unvented cathedral ceilings for seven years without any problem. Gary Nelson, an energy expert in Minneapolis, says that he has solved ice-dam problems in houses with cathedral ceilings by using a blower door and an infrared camera to find air leakage paths, then packing them full of cellulose.

In each case these folks are building a "warm roof"—one that has no vent space, as opposed to a "cold roof," which has cold air flowing through it. And in northern Canada, where wind-driven snow can infiltrate the best vents with frightening proficiency, warm roofs are becoming the norm, according to Rob Dumont, an engineer and research specialist with the National Research Council of Canada. The

success of a warm roof depends on a nearly perfect air/vapor barrier. But according to Ned Nisson in *Residential Building Design & Construction Workbook* (see suggested reading, p. 70), warm roofs are not a good idea where roof snow accumulations of 8 in. or more are common for extended periods of time because of potential ice-dam problems.

**Continuous soffit and ridge vents**—Despite some of the potential problems with roof ventilation, all engineers and scientists consulted for this article answered "yes" when asked directly, "Should I vent the roof?" And all agreed that the most efficient way to do it is with continuous soffit and continuous ridge vents.

Roof ventilation should be balanced between intake vents low in the roof and exhaust vents high in the roof. As long as there are no obstructions, air will move in a relatively straight line between the intake vents and the exhaust vents. Continuous soffit and ridge venting is the only system that directs this air uniformly along the underside of the entire roof. And this is where you want the ventilation, whether it's removing moisture, preventing ice dams or cooling the roof. Other types of exhaust vents, such as gable vents, roof louvers, and turbine vents, can leave portions of the roof unvented. This is true even when they're used in conjunction with soffit vents (which they should always be).

The 1:150 and 1:300 ratios are minimum requirements, developed for "leaky" houses. In general, continuous soffit and ridge vents provide three times the amount of NFVA required by the 1:300 ratio. Ideally, 50% of NFVA should be in the ridge, and 25% should be in the soffit on each side of the house.

Soffits are the best place for the intake vents because they're protected from the weather. And for maximum efficiency soffit vents should be as close to the fascia board as possible.

The most common complaint about ridge vents, especially among architects, is that they're ugly. But in a 32-ft. long house, you would need 11 roof louvers or 5 turbine vents to achieve the same NFVA as a continuous ridge vent.

Many architects and builders prefer the look of gable-end vents, and in some climates and situations they work fine. But gable vents are dependent on wind direction and speed to work. And when they work, gable vents move air down the center of the roof under the ridge, leaving large areas of the roof unvented. One thing you should never do is combine gable and ridge vents. This would result in short circuiting, where the ridge vent draws intake air from the gable vent rather than from the soffit vent.

Although there may be rare situations — shallow-pitched hip roofs, perhaps — where powered ventilation (attic and roof fans) is called for, powered ventilation generally costs more than it's worth. For one thing, power venting can depressurize the attic enough to suck moist air out of the house and into the attic, causing problems rather than solving them. Tests conducted by the Florida Solar Energy Center found that the additional attic air changes resulting from fan-forced ventila-tion did not cool the attic significantly enough to offset the cost of running the fan.

**Tricks and tricky spots** — Insulating over exterior walls while maintaining an air space for ventilation is a common problem. Two of the best solutions are either to use a raised-heel truss (drawing preceding page), or if you're framing with rafters, to run a plate across the tops of the ceiling joists and let the bird's mouth sit on it. Cardboard and polystyrene baffles are available for containing insulation over exterior walls and keeping it from blocking the soffit vents.

One of the easiest ways to encourage ventilation at the ridge is to drop the ridge board ¾ in. below the tops of the rafters (check this detail with your local building inspector first). This same idea works to move air over a hip rafter. Drop the hip rafter below the jacks, and let the sheathing bear on the jacks.

Dormers, especially if they have a cathedral ceiling in them, are tough to vent. If you can't use conventional methods, your best bet is to take special care with the air/vapor barrier around the dormer ceiling.

Venting around skylights in cathedral ceilings has always been a problem because the skylight blocks the air passage in one or more rafter bays. You can cut notches or drill holes in the cripple rafters above and below the skylight to provide an air channel into adjoining rafter bays. Some people will tell you this is a waste of time because ventilation air won't move laterally, at least not very well. Yet the same people agree that it doesn't hurt, and offer no alternatives.

The lack of ventilation around skylights makes them a common cause of ice dams. The folks who make Velux skylights (Velux-America Inc., P. O. Box 3268, Greenwood, S. C. 29648) recommend using an ice shield membrane on the roof deck 2 ft. to 3 ft. around the skylight and up the curb, under the flashing. This won't prevent ice dams, but it may keep the roof from leaking as a result of them.

One brand of skylight, CeeFlow Skylights (P. O. Box 98, Harbor Springs, Mich. 49740), actually boasts a vent space between its double-pane window and its acrylic lens. This allows an unbroken flow of air through the rafter bays that the skylight intersects. This may not eliminate ice dams, either, but sure seems a step in the right direction (for more on CeeFlow, see *FHB* #48 p. 90).

Kitchen and bathroom vent fans are a common source of attic moisture. They should *never* be vented into the attic or out the soffit. In fact, they should be located in walls rather than in ceilings and vented down and out through the floor. If the vent fan ducts do run through the attic or ceiling, insulate the ducts to keep condensation from accumulating inside them so that no moist air leaks into the cold roof.

Recessed lights are another problem area because most require that insulation be 3-in. away. This means that heat and air leak around them into the attic. If you can't avoid putting them in the ceiling, at least use the recessed can lights that can be insulated. They're called ICT lights and are made by various manufacturers, including Lightolier (100 Lighting Way, Secaucus, N. J. 07096) and Thomas Industries, Inc. (Residential Lighting Division, 950 Breckinridge Lane, Suite G50, Louisville, Ky. 40207).

**Cathedral ceilings** — The most common way to vent a cathedral ceiling built with dimension lumber is to use continuous soffit and continuous ridge vents and then maintain a 1-in. to 2-in. air space above the insulation in the rafter bays. The best way to maintain that air space is with polystyrene baffles stapled to the underside of the sheathing (photo left). These baffles, manufacturerd by various companies, prevent the fiberglass insulation from blocking the airspace. By framing the roof with scissors trusses, laminated wood I-beams or with parallel-chord trusses rather than solid lumber, you can obtain more room for insulation and ventilation.

Many timber-frame houses these days have cathedral ceilings built with stress-skin panels. To comply with the shingle warranty, timber-framer Tedd Benson nails furring strips horizontally and a layer of plywood on top of the panels as a nail base for asphalt shingles. Then he vents the space by dropping the crown mold along the rakes slightly and covering the space with screening.

Although Benson doesn't use them, another approach would be to cover the roof with panels that have air spaces integrated into their design. Both Branch River Foam Plastics, Inc. (15 Thurbers Blvd., Smithfield, R. I. 02917) and Cornell Corp. (P. O. Box 338, Cornell, Wisc. 54732) make such panels.

**Manufactured vents** — Many companies make soffit and ridge vents, and whose products you use will likely depend more on what's available from your favorite supplier than on which product performs best. Air Vent, Inc. (4801 N. Prospect Rd., Peoria Heights, Il. 61614) offers a number of products that represent what's available (photos facing page).

Air Vent's basic ridge vent is made of aluminum (.019 in. thick) and has an integral baffle, which they claim reduces rain and snow infiltration and enhances ventilation by deflecting wind over the vent and creating an area of negative pressure (the Venturi effect) that pumps air out of the attic. Air Vent also sells the same vent with a fiberglass filter in it (Ridge Filtervent) that's supposed to reduce further the chance of rain or snow infilatration. According to Air Vent, the filter also deters insects, particularly cockroaches (a problem in warm climates), from entering the attic.

Air Vent makes a product called Flash Filtervent that allows ventilation where a shed roof meets a vertical wall, and one called Utility Vent that is simply an integral baffle and louver you can use to make your own ridge vent. Air Vent sells a shingle-over ridge vent (Shinglevent), also with an integral baffle, made of high-density polyethylene. A shingle-over

**The most common way to vent a cathedral ceiling is with polystyrene baffles stapled against the roof sheathing to prevent the fiberglass batts from blocking the air path in the rafter bay.**

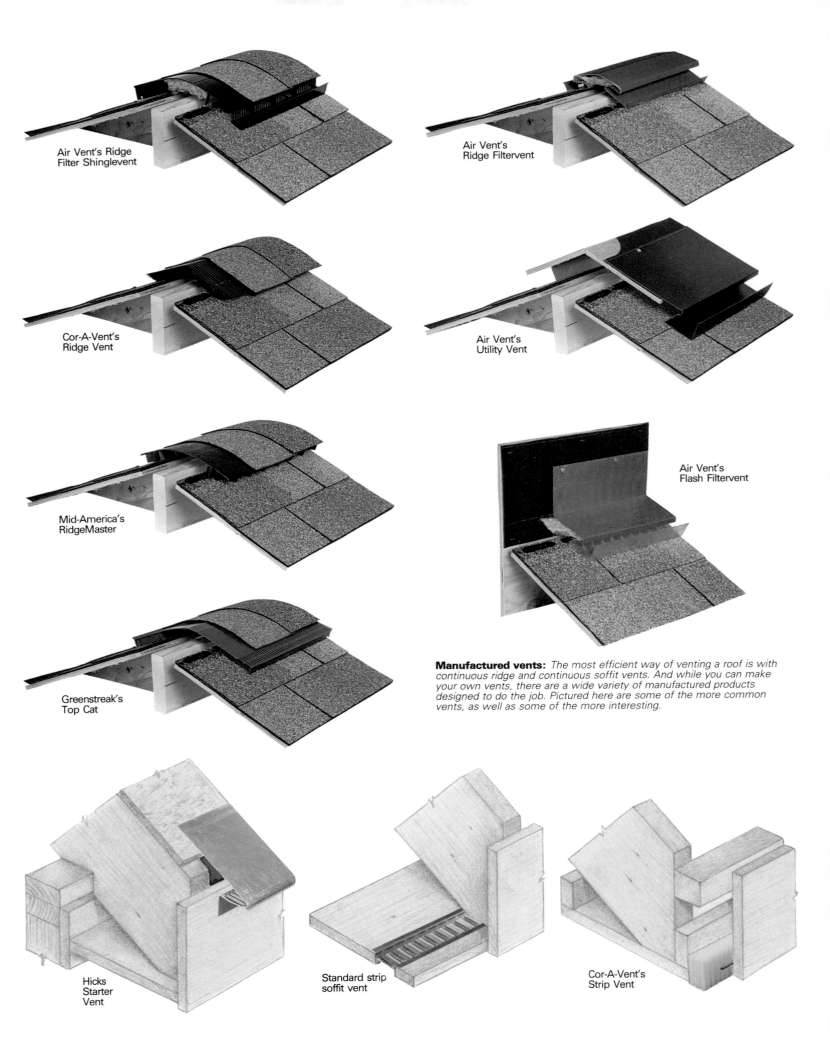

Air Vent's Ridge
Filter Shinglevent

Air Vent's
Ridge Filtervent

Cor-A-Vent's
Ridge Vent

Air Vent's
Utility Vent

Mid-America's
RidgeMaster

Air Vent's
Flash Filtervent

Greenstreak's
Top Cat

**Manufactured vents:** *The most efficient way of venting a roof is with continuous ridge and continuous soffit vents. And while you can make your own vents, there are a wide variety of manufactured products designed to do the job. Pictured here are some of the more common vents, as well as some of the more interesting.*

Hicks
Starter
Vent

Standard strip
soffit vent

Cor-A-Vent's
Strip Vent

vent is a ridge vent over which you can nail ridge caps to make the vent less obtrusive.

The original shingle-over vent was Cor-A-Vent (Cor-A-Vent, Inc., 16250 Petro Drive, Mishawaka, Ind. 46544). It was invented in the early 1970s by Gary Sells, a builder in northern Indiana. The prototype was made with layers of corrugated cardboard, coated with epoxy, and according to Sells, is still in place today. Sells had to wait for plastic technology to

catch up with him before he could go into production. Now Cor-A-Vent is made of six layers of corrugated plastic, stapled together and beveled (photo previous page). By cutting Cor-A-Vent in half lengthwise, you can use it to vent shed roofs and clerestories. Cor-A-Vent also makes a 1½-in. by 1-in. corrugated plastic soffit vent, available in white or black.

At the same time that Sells was developing his product, another builder, Robert Hicks,

was designing a combination drip edge and vent (photo previous page). Since then several other companies have come out with similar products, but Hicks Starter Vent (124 Main St., Westford, Mass. 01886) was the first. Drip-edge vent is installed the same way as standard drip edge, but extends beyond the roof sheathing to create a sheltered overhang for a row of louvers. Besides their use on houses with no soffits, drip-edge vents are often the easiest way to retrofit intake vents on a house.

Over the past few years many new ridge vents have appeared on the market, including RidgeMaster (Mid-America Building Products Corp., 9246 Hubbell Ave., Detroit, Mich. 48228), a polypropylene vent with a system of internal baffles and a layer of filter material woven through them (photo previous page). Another polypropylene vent, Top Cat (Greenstreak, 3400 Tree Court Industrial Blvd., St. Louis, Mo. 63122) has vent openings in the top rather than along the sides (photo previous page). Unfortunately, there is no trade association of passive-vent manufacturers, so it's tough to know how many others might be out there.

**Site-built vents**—A lot of builders make their own soffit and ridge vents because they don't like the look of manufactured vents. The drawing at the left shows one design for building a ridge vent (other designs were featured in *FHB* #21 pp. 51-53 and *FHB* #52 p. 50). To create their own soffit vents, builders usually cut a slot in the soffit and staple screening to the backside. It is also possible to cut decorative shapes in the soffit. Some builders use a 2-piece fascia board with an air space between, covered with screening. It's not advisable to use insect screen (1/16-in. mesh) to make vents, though. Insect screen is 50% closed, meaning it reduces by 50% the NFVA of whatever opening it covers. Also, insect screen can easily become clogged with dirt or paint. Use 1/8-in. mesh or larger. □

*Kevin Ireton is managing editor of* Fine Homebuilding.

**Suggested reading**
Here are some of the publications I found helpful while working on this article.   — *K. I.*
• *Principles of Attic Ventilation,* available from Air Vent, Inc., 4801 N. Prospect Rd., Peoria Heights, Il. 61614. $3, 30 pp. *Probably the best single source of information on the mechanics of attic ventilation.*
• *Roof-Snow Behavior and Ice-Dam Prevention in Residential Housing,* by Howard L. Grange and Lewis T. Hendricks, available free from Air Vent. *A thorough treatment of ice dams, their causes and solutions.*
• *Residential Building Design & Construction Workbook,* by J. D. Ned Nisson, Cutter Information Corp., 37 Broadway, Arlington, Mass. 02174, 1988. $85, softcover, 347 pp. *Expensive, but worth the cost. An in-depth explanation of superinsulation theories and techniques, including information about airtight construction and roof ventilation.*

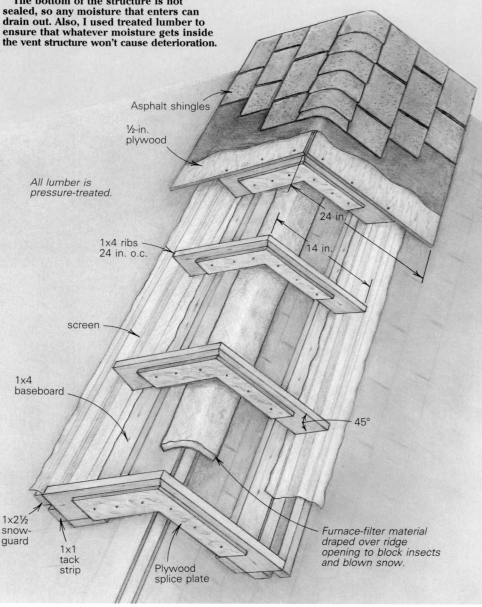

### The permanent ridge vent

During the construction of our cathedral-ceiling family room, I installed a prefabricated aluminum ridge vent. Three years later it started leaking, and rather than trying to solve the leaks, I decided to install a permanent, maintenance-free vent structure.

I used the same construction as the basic roof—plywood sheathing covered with asphalt shingles. It was easy to build, blended with the existing roof and the only maintenance will be the replacement of shingles every 20 years, along with the those on the rest of the house.

The bottom of the structure is not sealed, so any moisture that enters can drain out. Also, I used treated lumber to ensure that whatever moisture gets inside the vent structure won't cause deterioration.

I needed an overall length of 32 ft., so I opted for two 6-ft. and two 10-ft. sections. This way the 8-ft. lengths of plywood overlap the joints between the frame sections and tie the whole structure together. I assembled each section in two halves, working on the floor of my garage. Once the frame was in place on the ridge, the sheathing pieces were glued and screwed to the ribs to resist uplift forces during high winds.
—*R. W. Missell, an electrical engineer in Annandale, Va.*

Asphalt shingles

½-in. plywood

All lumber is pressure-treated.

24 in.

1x4 ribs 24 in. o.c.

14 in.

screen

1x4 baseboard

45°

1x2½ snow-guard

1x1 tack strip

Plywood splice plate

Furnace-filter material draped over ridge opening to block insects and blown snow.

# Built-Up Cedar Roofing
## Common shingles mimick a traditional reed thatch

by Steve Dunleavy

**A**bout eight years ago my company was hired to put a roof on a custom home in the Sierra Nevada mountains overlooking Lake Tahoe. The house was a massive Tudor mansion with a steep roof, incorporating turrets, dormers, hips and valleys. The builder had spent a lot of time researching European architecture and wanted the house to look authentic, including the roof, which was to be reed thatching.

True thatch roofs are practically non-existent in North America. Importing material and labor, primarily from England, has proven to be prohibitively expensive— most companies who have tried it have met with failure. Building departments and insurance companies unfamiliar with what amounts to a Middle Ages technology have been hesitant to approve or underwrite what my dictionary calls "a roof covering made of straw, reeds, leaves, rushes, etc." Also, I've seen thatch roofs play host to every sort of insect, bird, rodent and rot imaginable. For all these reasons, the reed-thatch plan was scuttled and I was asked to simulate, using cedar shingles, the droopy, layered look of authentic thatch roofing. Beginning with that first house and refining and borrowing elements from many styles of shingling, we devised a shingle roof that we call "built-up cedar thatching." My company has now installed more than two dozen of them, and they've proven to be both durable and dazzling to the eye, capturing the velvety look of traditional reed thatch (top photo).

**Division of labor—**From my standpoint, applying built-up cedar thatching seems more like a military campaign than a roofing job. Everything involved is on a large scale: the intense labor; the massive amounts of material stored and then moved higher and higher up steep inclines; the wear and tear on men, tools and equipment. It's no surprise that labor is the biggest expense with these roofs. Labor costs must be budgeted carefully based on the job, the climate and the topography of the roof itself.

I break my crews into two categories: nailers and laborers (or grunts and sub-grunts, as

Though it resembles a traditional reed thatch, a built-up shingle roof consists of cedar shingles split to a maximum width of 5 in. and laid in random patterns with exposures of 3 in. or less. In the top photo the shingles wrap around radiused rakes and eaves. At hips (photo above), shingles are tapered with a utility knife and laid like pie slices. The angle of the taper depends on the roof pitch. A layer of Ice & Water Shield beneath the roofing felt at both hips and valleys adds protection against the weather.

they've graciously named each other). The nailers are usually journeymen roofers who are familiar with built-up shingle thatching (although experienced laborers occasionally nail, too). Nailers are also responsible for installing weatherproof membranes, roofing felt and metal flashings.

The laborers' job is to keep the nailers supplied with shingles, which is more difficult than it sounds. A built-up shingle thatch requires two to three times the amount of material needed on a standard shingle roof, and every shingle has to be hauled up to the roof via winch or ladder as the job progresses (there's just too much material for a rooftop delivery). Also, laborers are responsible for splitting each shingle to a width of 5 in. or less, which is critical to the appearance of these roofs.

If the roof framing is curved at the eaves, it's usually necessary to steam-bend shingles prior to nailing them up. This is a full-time job for one man (more on that later).

**Everyday materials—**While a built-up shingle thatch might appear exotic, the materials we use to assemble it are not. We prefer 16-in. long, #1-grade Western red cedar shingles, which are 100% edge grain and clear heartwood. We've tried using 18-in. long shingles, but found them to be harder to split and steam-bend, as well as a little too thick to lie properly. We've had some success with #2-grade shingles, which are clear and knot-free for a minimum 10 in. from the butts. They cost $10 to $20 less per square than #1-grade shingles, so they're occasionally specified by clients to save money. These shingles don't affect the appearance of a roof, but they can be difficult to split and bend because of knots, leading to increased waste and more work. Generally, I figure that a simple cedar thatch roof will be at least three times the cost of a standard roof; complicated roofs run even more. Labor is easily 50% of the job.

Because of ice-dam problems in snow country, we're required by code to install an elastic-sheet membrane over the first nine feet up from the eaves. We use Ice & Water Shield (W. R. Grace and Co., 62 Whittemore Ave.,

Cambridge, Mass. 02140), which comes in 3-ft. wide rolls that cover 225 sq. feet. The product consists of rubberized asphalt with a cross-laminated polyethylene film on one side and a pressure-sensitive adhesive backed by removable kraft paper on the other. The sheeting is installed polyethylene side up, forming an impermeable membrane that stays watertight even when nails or staples are driven through it. Ice & Water Shield has an embossed surface, so it's not slippery to walk on, an endearing feature to roofers.

In addition to using the membrane along eaves, we use it as an extra precaution under the flashing at roof-wall intersections, valleys, chimney saddles and other potential trouble spots. At about $50 per square it's expensive, but well worth its price for the protection it offers. A warning, though: Ice & Water Shield is the stickiest flypaper ever invented. On a steep roof it's a two man operation to put down, and we don't mess with it in any sort of wind.

We are also required to install tar paper over the membrane. We use 30-lb. felt affixed to the deck with barbed roofing nails. The felt is substantial enough to provide temporary weatherproofing and serves as a durable substrate for a built-up thatch.

For flashings we prefer copper, although many people ask for galvanized sheet metal to save money. We fabricate and install most of our own flashings, but if a peculiar bend or solder job is giving us trouble, we call in the experts. The quality of the flashing material and workmanship is critical because these roofs, if installed properly, should last a century or more. There are times when nothing else will work except a tube of gun-grade urethane caulk, so we always have some available in case the need arises.

**Tooling up—**We prefer to use pneumatic staplers to apply shingles, as does most of the roofing industry. Indeed, roofs like this, typically requiring a quarter million or more fasteners, would be astronomically expensive and virtually impossible to apply if they had to be hand nailed. Our weapon of choice is the Senco MII pneumatic staple gun (Senco Products Inc., 8485 Broadwell Rd., Cincinnati, Ohio 45244), loaded with 2-in. staples. These tools are almost indestructible on the outside (every one I own has cartwheeled off more roofs than my esteemed employees care to admit). When the insides go, virtually everything can be replaced using inexpensive repair kits. I've rebuilt these guns on the tailgate of my truck at the job site. Staples can be bought almost anywhere, either generic or brand name. We use Senco brand staples because the tolerances seem a little tighter and they don't seem to jam or break the tools as often as other staples do. Pneumatic tools are percussive and loud, so we wear disposable foam earplugs, which can dramatically reduce fatigue.

We use a variety of saws in our thatch work. For on-the-roof cutting of shingles we prefer small circular saws, such as the Makita 4200N. Detail work around skylights and chimneys re-

**Shingles for radiused eaves are steamed and bent to the proper curvature before installation. The author's steamer consists of a truck utility box modified to fit over a 55-gal. drum of boiling water (top photo). Once shingles are saturated with moisture and become pliable, they're transferred to the bending jig (middle photo). When the jig's lid is closed, a pressure bar in the lid forces the shingles to bend over a platen to the proper radius (bottom photo). Typically, about 10% of the shingles break while bending.**

quires plenty of cuts, and these lightweight trim saws are more comfortable and safer to use than are the larger 7¼-in. circular saws (for more on trim saws, see *FHB* #48, pp. 40-43). We use a four-tooth carbide blade, which doesn't make a really precise cut but will last the length of the job (we're not, after all, building a piano here). This combination of high-rpm saw and four-tooth blade is effective but potentially dangerous because of the size and velocity of the waste it expels (the waste is more akin to chain-saw chips than sawdust). Eye protection is a must.

Cutting through several layers of shingles, such as at the ridge of a roof, requires more power than a trim saw offers, so we switch to

a standard 7¼-in. worm-drive saw. Weaving a built-up thatch over hips, curved rakes and other obstacles can require thousands of shingles to be cut at predetermined angles. We accomplish this with a radial-arm saw set up on the ground, and haul tapered shingles up on the roof as we need them.

**Bending shingles—**The steam boxes and bending jigs we use change from job to job, depending on what's needed. Our steamer consists of an old truck utility box with a metal flange welded on the bottom that fits over a 55-gal. drum full of boiling water (top photo). Water is heated by a propane weed burner at the base of the drum, and a large opening in the bottom of the box allows steam to enter. We load and unload material through the hinged door; a 2x4 rack with a row of nails front and back holds the shingles on edge. A second rack can be stacked on top, allowing the box to hold twice as many shingles. We rotate our stock, replacing hot, pliable shingles with fresh ones. The idea is to be as efficient as possible—we don't leave the steambox door open too long, and we have the bending jig open and ready for shingles when they come out of the steamer. Shingles are usually steamed for 20 to 30 minutes. There's some springback, but not much.

Our bending jig comes from the same make-it-on-the-spot family as the steamer. It can be made out of scrap lumber in a few hours (middle photo). The jig is simply a hinged frame made of 2xs and plywood, with a strategically placed 2x2 bending platen in the base of the frame and a 2x4 pressure bar in the lid. Shingles are loaded into the base butt-end down, with the butts constrained by a 1x4 crossbar. When the lid is closed, the pressure bar forces the shingles into a curvature dictated by the relative positions of the pressure bar and bending platen (bottom photo). We usually have to fine-tune the jig to achieve a particular bending radius, and we figure on breaking about 10% of the shingles.

The jig is screwed together with "grabbers" (similar to drywall screws), both for strength and so it can be adjusted. We intentionally build our jigs small so that we can transfer shingles quickly from the steamer to the bending jig before they have a chance to cool off.

**Thatching guidelines—**The most important factors for achieving the look of a thatched roof are shingle size, exposure and pattern. A typical non-thatch shingle roof is composed of straight 5-in. courses made up of shingles 4 in. to 14 in. wide, and measures about ⅞ in. thick. A built-up thatch roof, on the other hand, is composed of smaller shingles (5 in. wide or less) laid in a random pattern with exposures ranging from ½ in. to 3 in. The result is a roof surface that's about 3 in. thick.

Breaking shingles to the proper width is more complicated than it sounds. We break them by hand because it's quicker and safer than swinging a hatchet thousands of times.

By grasping a shingle on its side edges with thumbs in the middle and quickly snapping the wrists, we usually get a clean break. Moist shingles can bend almost in half before breaking and require more effort, as do thick or dry shingles. Beginners tend to get sore hands quickly and usually try to break material over their knee caps, then get sore knees and go back to the first method. Sometimes the grain isn't straight, resulting in an angled break. These pieces are set aside to be trimmed later with a knife. Breaking wide material into smaller pieces goes against every good roofer's natural instincts, but narrow shingles are an important component of the built-up thatch look.

While there are no discernible rows in a built-up shingle thatch as opposed to a standard shingle roof, many of the same installation rules apply: all eaves have doubled, sometimes tripled courses; each shingle is fastened with two galvanized staples ¾ in. from the side edges and about 6 in. up from the butts; adjacent edges are gapped, although not as much as the ⅜ in. sometimes recommended; and joints between adjacent courses are usually offset by at least one inch (we bend this rule a little because of the number of courses in a built-up shingle thatch).

Almost all shingles are installed plumb, with butts square to the vertical fall line of the roof. This is more difficult than it sounds. On a steep pitch, material can be nailed mistakenly at a slight angle because the roofer is concentrating on a relatively small roof section, without the benefit of well-defined rows. Occasionally a nailer will have to scrutinize his work to regain an idea of what's "up" and "down."

The random pattern we apply in our thatching isn't as casual as it might appear. Many different nailers can be working on a project at once, and it's important that their work match seamlessly. From observing each other's shingling, nailers usually develop a sort of "mind meld," where individual differences in workmanship become impossible to detect.

In the last few years we've seen a growing interest in curved roofs, with architects and builders striving to design and construct eye-catching houses with radiused rakes and eaves and with virtually no joints at intersecting roof planes. As far as I can tell, there is no standard way to frame these curves (for one approach, see *FHB #23, pp. 52-56*). As for shingling them, if some crazed carpenter wants to jelly-roll sheets of plywood until his roofline looks like a roller coaster, with enough head-scratching we'll figure out a way to shingle it.

**Eaves, gables, hips and valleys—**The procedure for shingling eaves is basically the same whether they're square or radiused. For square eaves, we start by applying a double or triple course of shingles (depending on the desired appearance) over the elastic membrane and felt, making sure to offset the joints. If the eave is finished with a fascia board, we install our shingles with a mini-

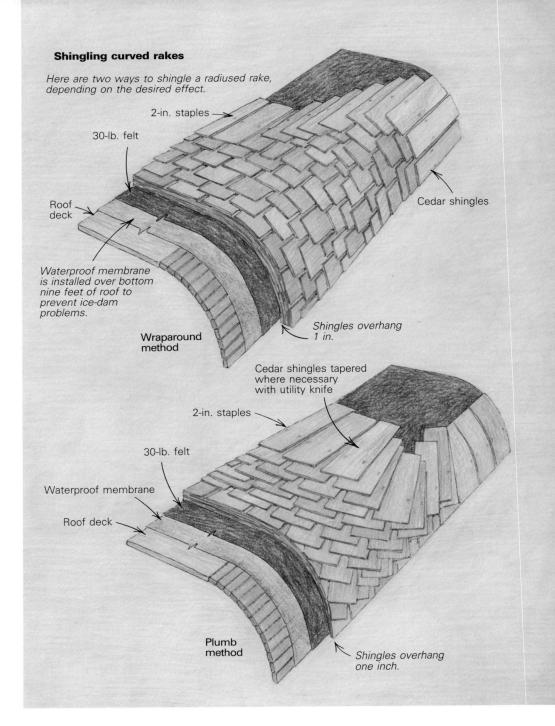

**Shingling curved rakes**

*Here are two ways to shingle a radiused rake, depending on the desired effect.*

2-in. staples

30-lb. felt

Roof deck

*Waterproof membrane is installed over bottom nine feet of roof to prevent ice-dam problems.*

Cedar shingles

**Wraparound method**

*Shingles overhang 1 in.*

Cedar shingles tapered where necessary with utility knife

2-in. staples

30-lb. felt

Waterproof membrane

Roof deck

**Plumb method**

*Shingles overhang one inch.*

mum 1½-in. overhang. If there is no fascia board—if a builder prefers, say, plaster soffits butted to the shingles—we allow more of an overhang to accommodate furring strips, lath and stucco (soffits are recommended because of the large quantity of staples penetrating the overhangs). Radiused eaves require the use of shingles pre-bent to the same radius as the eave. The number of curved shingles is dependent on the pitch of the roof (the greater the pitch, the fewer the number of curved shingles needed).

To enhance either radiused or square eaves, we sometimes install the starting course of shingles so that it undulates along the edge of a roof (top photo, p. 75). We do this by tapering the shingles with a utility knife and by "wave-coursing" the shingles in convex and concave arcs, making sure the apex of the arc overhangs the roof a minimum of 1½

in. The maximum overhang winds up being about five inches, which is possible because of the added strength and rigidity afforded by the small exposures. Usually builders don't want these arcs to be uniform, preferring an edge that appears to sag or droop in spots, so we lay the shingles free-style, with no layout beforehand. After this first decorative row is installed, we fill in the low spots with the usual random coursing.

Gable ends are also either square or radiused. On square rakes shingles are applied in the usual way, with a 1-in. overhang. For curved rakes, shingles are often wrapped around the bend of the roof, with the edges of the shingles parallel to the edge of the rake (top drawing above). Unless the radius of the framing is especially tight, the use of pre-bent shingles is unnecessary (remember we're dealing with narrow shingles that divide the radius down into small increments). The shingles are laid with a 1-in. overhang, the same as for a flat gable end. This is a fast, efficient method of shingling a curved rake, and we've done many of them this way. However, it's never appeared quite right to me because instead of the shingles being plumb along the radius, they're at an angle.

I prefer to fan the shingles down from the roof decking onto the radiused section so that the shingles remain somewhat plumb (bottom drawing, p. 73). This requires that the shingles making up the fan be tapered, that the bottoms of the shingles be cut at an angle along the edge of the roof to match the roof pitch and, finally, that a starter course be installed along the edge of the rake. It's a lot of extra effort, but I think it's worthwhile. To really confound matters, we sometimes lay the courses at the edge so that they undu-

late, too. In this case, we use the same procedure outlined above for the eave.

Hips and valleys, curved or not, are less complicated than curved gable ends. By weaving the shingles, we create the image of one roof section gently flowing into another. For hips, we accomplish this by tapering individual shingles to a particular angle (much like we do for arcs and curved rakes) and then combining them like pie slices until the proper turn has been made (bottom photo, p. 71). The angle of the taper depends on the roof pitch. For valleys, we simply open up the joints between adjacent shingles so that the shingles touch only at their butt ends (middle photo, facing page). Subsequent courses then cover the open joints. For both hips and valleys, a layer of Ice & Water Shield beneath the shingles serves as extra protection against moisture penetration. These same techniques allow us to roll a built-up shingle roof over virtually any rooftop structure, including all types of dormers and turrets.

**Capping the ridge**—For ridge caps, we use concrete barrel tiles, commercial cedar ridge cap or copper flashing (drawings left). Concrete tiles can be found in a variety of shapes and colors. We fasten them with 20d nails to allow for the thickness of the tile and the shingles, and we fill the gaps under and between the lapped tiles with mortar dyed to the appropriate color. While applying mortar, we cover adjacent shingles with polyethylene sheeting to protect them from staining.

Cedar ridge cap is made in lengths of 16 in. and 24 in. (we prefer 24 in.). This type of ridge tends to get lost on a massive thatch roof, so we recommend that adjacent lengths be lapped with about 5 in. to the weather (about one half the normal exposure), with the first course doubled on each end of the roof. This doubles the amount of ridge cap needed, but creates a bold, distinctive ridgeline. We fasten this cap with stainless-steel or galvanized 10d nails to assure adequate penetration into the sheathing.

Pre-bent copper (30 ga. or 28 ga.) also works well with a built-up cedar thatch, though it's more expensive then cedar. Our favorite design uses 10-ft. long sections bent in an "arrowhead" profile and fastened directly to the sheathing prior to shingling. Joints are lapped a minimum 4 in. and are caulked or soldered. The top few courses of shingles on either side are then cut progressively shorter so that their tops butt against the barb of the arrowhead. A bead of polyurethane sealant at the joint between the shingles and flashing prevents water from running under the shingles.

**Flashing**—Clients don't usually want to see lots of flashing on their roofs, so we're constantly weighing aesthetics against overall durability. Our real purpose as roofers is, after all, to make dwellings impervious to weather.

For plumbing and HVAC penetrations, the galvanized roof flashings usually supplied by subcontractors are acceptable. We install

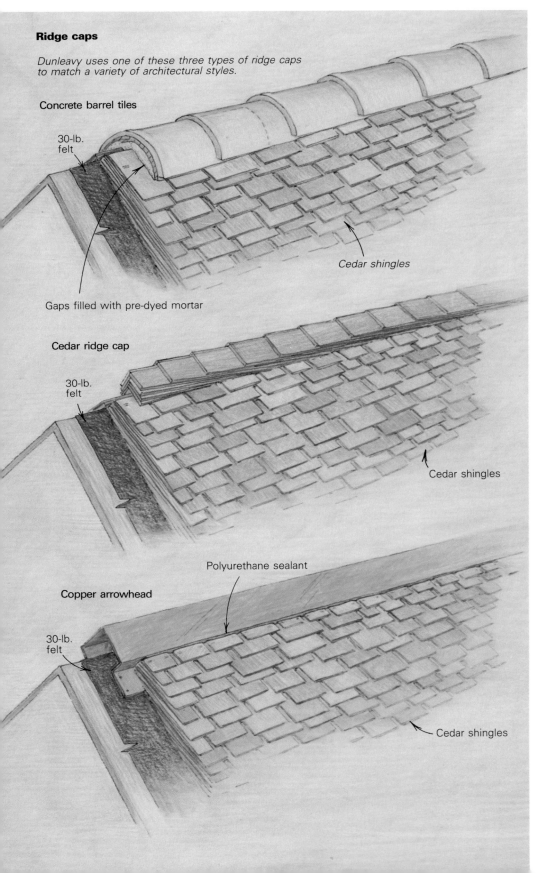

**Ridge caps**

*Dunleavy uses one of these three types of ridge caps to match a variety of architectural styles.*

**Concrete barrel tiles**

30-lb. felt

Gaps filled with pre-dyed mortar

*Cedar shingles*

**Cedar ridge cap**

30-lb. felt

*Cedar shingles*

**Copper arrowhead**

Polyurethane sealant

30-lb. felt

*Cedar shingles*

Instead of radiused rakes and eaves, some roofs terminate at square roof edges. As shown in the photo above, wave-coursing was used for the starter courses to create an authentic droopy appearance.

them much as we would on a standard-shingle roof. For sidewall flashing, we prefer a continuous piece of pre-bent copper or galvanized sheet metal over Ice & Water Shield and 30-lb. felt. We fasten each shingle with the staples placed as far away as possible from the flashing and maintain a 1-in. wide channel between the shingles and the wall to allow for water run-off. We avoid using step flashing because it's almost impossible to install properly on our built-up shingle roofs.

Because there are so many thicknesses of shingles in a built-up shingle thatch, we can hide the base flashing of a wall, skylight or chimney and still get a waterproof joint (bottom photo). To accomplish this, we run three or four full-length courses of shingles under the flashing and three or four progressively shorter courses over it. We're careful to place the staples toward the butts and at an angle to avoid penetrating the flashing. A high-quality caulk or flanged counterflashing is used to seal the joint where the shingles butt into the flashing. At the corners, we leave a small gap between the shingles. This method allows for drainage around any flashed component.

Roof-to-masonry joints can be handled the same way, although at times inappropriate counterflashing has been installed prior to our starting a project, forcing us to improvise. A classic example was a granite chimney and built-up cedar thatch, both intended to last a century or more, flashed with a thin piece of roll aluminum that might last 10 years.

Thatch roofs are closer to stonework than to roofing and hark back to an era in which a craftsman's work might well outlast him. As always, planning is the name of the game. □

*Steve Dunleavy is a roofing contractor and owner of Hi & Dry Roofing in Stateline, Nevada. Photos by author except where noted.*

As with hips, shingles are installed over valleys in continuous courses, creating a smooth transition between adjacent roof sections. Valley shingles are fanned out so that the shingles touch only at their butt ends. Shingles are affixed with the aid of pneumatic staplers.

**The multiple layers of shingles in a built-up shingle thatch roof allow base flashings for skylights, walls and chimneys to lie hidden beneath two or three shingle courses, creating a more pleasing appearance. For the skylight pictured at right, a continuous bead of urethane caulk between the tops of the shingles and the base will keep water out. The side flashings are continuous because of the difficulty of step-flashing this type of roof, which is typically eight or nine courses thick. Water runoff is encouraged by holding shingles an inch away from the sides of the skylight, creating a gap between the shingles at the bottom corners.**

# Roofing with Slate

## Shingling with stone will give you a roof that lasts for generations

by David Heim

When my wife Katie and I bought a farmhouse in northeastern Pennsylvania a few years ago, we weren't surprised that it had a slate roof. Built sometime before 1860, the house is close to several slate quarries. Here and in other parts of the Northeast (photos below), slate roofs are common on old houses.

After more than 120 years, the original slate on our roof was suffering from age and neglect. But it was one of the most attractive features of the house, so our plan from the outset

was to repair or replace it. The roof didn't leak when we bought the house, although it had in the past. Previous owners had smeared tar in between the slates in several places, and some of the slates had been replaced with pieces of tin. Many slates were soft to the touch and crumbled readily.

Over time, even slate yields to the intrusion of water; and once this happens, freeze-thaw cycling causes it to delaminate along its cleavage planes (sidebar, facing page). Cracks in

the surface (called crazing), flaking (spalling) and chalky deposits around the edges and unexposed face of the slate are signs that slate has reached the end of its useful lifespan.

We could have patched the roof to make it last a few more years, but we decided instead to add a new roof to the list of improvements we had planned for the house. Our first inquiries into re-roofing with slate didn't bring very positive responses. Slate is prohibitively expensive, we were told; it's too hard to work

with, and there aren't any good slate roofers around any more.

After some persistent investigating, though, we realized that what we'd been told wasn't entirely true. In my area, slate isn't much more expensive than other good roofing materials—particularly not in the long run. And we found a contractor, Jim Hilgert, who knows slate work backwards and forwards. Having seen how Hilgert's crew handled our roof, I'm convinced that roofing with slate isn't much harder than roofing with other shingles. It's in cutting, hole punching and handling that slate work differs. Though the job shown here is a re-roof, you would use the same tools and techniques to put on a new slate roof.

**Selecting the material**—Slate comes in a wide variety of colors, depending on where it is quarried. For our re-roofing job, Hilgert used #1 clear Pennsylvania blue-grey slate, salvaged from a barn in New Jersey that was about to be torn down. The pieces were 24 in. by 12 in. (with an exposure of 10½ in.), the same size as our originals. The slate was in good condition and only about 20 years old—

not yet middle-aged, as slate goes. Pennsylvania slate is reputed to last for at least 75 years. Slate quarried in Vermont and Virginia can last 150 years or more on a roof.

Including removal and transport, we paid $100 a square (enough to cover 100 sq. ft.) for our slate, or about half the local quarry price at the time. Vermont slate can sell for up to $300 a square, but even at that price it's not a bad deal when you consider the longevity of the material. Fiberglass shingles, which cost about $60 a square in our area (including roof sheathing and felt underlayment), would have to be replaced three or more times within the lifespan of a single slate roof.

Recycled slate isn't always the bargain it appears to be. Usable old material can often be impossible to find, even in an area where slate is commonly used. For example, four months after Hilgert bought our slate, he was unable to find more second-hand slate for another job. When you do come across salvageable material, it takes time and experience to cull out the bad slate.

To be sure that none of our slates had hairline cracks or delaminations, Hilgert "rang"

the slates as he pulled them off the barn roof, much as you'd ring a china cup to check its soundness. If you hear a faint echo when you tap the slate—something like the sound your knuckles make when they rap a solid plank of wood—the slate is all right. A dull thud with no resonance indicates unsound stone that's best rejected. Hilgert also rejected "ribbon" slates—pieces that have a pale streak running through them. This impurity in the stone is a weak spot that won't weather well and will crack prematurely. Slates without ribbons are called "clear."

Once he'd found enough material, Hilgert hosed the slates down to remove the accumulated stone dust. Dipping salvaged slate in a solution of oxalic acid and water (wear rubber gloves) will remove weathering marks and restore the slate surface to good-as-new condition, but it takes a lot of time. Hilgert trucked the cleaned slates to our house and stacked them on edge, like large, thin dominoes.

**Preparing the framing**—Like many other houses in our area, our roof has almost no sheathing. The slates were fastened to roof

## From quarry to roof

In terms of composition, slate is little different from the clay deposits you might find in a river bed. It's the geological forces of pressure, temperature and time that transform clay into shale and slate. Both are sedimentary rocks, but shale is softer and less dense because it hasn't been cooked or compressed as much as slate. When slate forms, tremendous temperatures and pressures cause the mineral grains to align so that they're parallel to each other. This granular alignment creates the cleavage planes that enable quarry workers to split out thin, flat sheets of stone.

Splitting slate along its cleavage plane reveals the surface texture, or grain, of the slate. On premium-quality slates, the grain should run lengthwise, as it does on a cedar shingle. Grain can vary from smooth to coarse, and a rough surface doesn't mean that the slate is poor-quality material. Smooth slates are easier to work with, however.

Slate color depends on chemical and mineral makeup, and can vary from the grey stone quarried in eastern Pennsylvania to the red and green tones found along the Vermont-New York border. Other standard colors established by the Department of Commerce are black, blue-black, blue-grey, purple, mottled green and purple. *Ribbon* slates are streaked because of impurities in the original clay deposit. In some cases, this ribbon can weather prematurely, so slates classified as *clear* are a safer bet for a long-lasting roof. Color is further

qualified as either *unfading* or *weathering*. Some slates change color over time, but those designated as unfading will not.

Standard roofing slate is ³⁄₁₆ in. thick and can be ordered in a number of sizes, from 10 in. by 6 in. to 24 in. by 14 in. These slates are fairly uniform and usually have their holes (two per slate) machine-punched at the quarry. To install what is known as a *textured* slate roof, you'd use slates that vary in thickness from ³⁄₁₆ in. to ³⁄₈ in. The *graduated* slate roof is another variation involving slates of different sizes and thicknesses. Usually the larger, thicker (sometimes up to 2 in.) slates are located near the eaves, with thinner slates and less exposure used near the ridge. These roofs allow considerable aesthetic expression on the part of the slater, and no two are the same, as the photos at left, taken in New England, show. Most of the slate work done today, however, is with standard slates.

*Ordering slates*—Like other roofing materials, slates are sold by the square. A square of slates should cover 100 sq. ft., with the standard 3-in. lap. Slate size determines the number of slates in a square, and the exposure to the weather. Exposure is easily figured with a simple formula: Subtract 3 in. from the length of the slate, and divide by 2. The 24-in. by 12-in. slates used for Heim's roof come 115 to the square; 12-in. by 8-in. slates come 400 to the square.

Slate prices can vary a great

deal, depending on size, thickness and color. Quarry prices start at $300 to $400 per square. Unless you're near a supplier (see the list of operating quarries below), freight charges may end up determining what your best delivered price is. Most quarries don't have a full range of sizes in stock. Special orders can be cut, but you'll have to wait for them. And remember that the smaller size slate you use, the longer it will take to nail up. Larger slates—18 in. or longer in standard or random widths—can really go up quickly. What this boils down to is that a little phone work can go a long way toward saving time and money.

If you're new to slate roofing, there's a good book available from Vermont Structural Slate Co., Inc. (Box 98, Fair Haven, Vt. 05743; $7.95 postpaid). Entitled *Slate Roofs* and originally published in 1926, the book provides a detailed, state-of-the-art look at slate work in its heyday.

Below are names, addresses and telephone numbers of four major slate quarries that operate on a year-round basis.

Buckingham Virginia Slate Corp., Box 11002, 4110 Fitzhugh Ave., Richmond, Va. 23230; (804) 355-4351.

Rising and Nelson Slate Co., West Pawlet, Vt. 05775; (802) 645-0150.

Structural Slate Co., 222 E. Main St., Pen Argyl, Pa. 18072; (215) 863-4141.

Vermont Structural Slate Co., Inc.; Box 98, Fair Haven, Vt. 05743; (802) 265-4933.   —*Tim Snyder*

From *Fine Homebuilding* (April 1984) 20:38-43

**Traditional tools.** The slater's stake is T-shaped, and its sharp end can be driven into a rafter or other wood work surface. Its horizontal edge supports the slate while it's punched, cut and smoothed. The hammer, which is made from a single piece of drop-forged steel, is designed to drive and pull roofing nails, to punch holes and smooth the rough edges of cut slate.

laths (purlins) of 4/4 by 2-in. hemlock that had been nailed across roughsawn 4x6 rafters.

The roof lath on our house is 10½ in. o. c., the spacing required to give our 24-in. by 12-in. slates a 3-in. lap (drawing, facing page). (Lap refers to the required triple overlay of slates on three consecutive courses.) Roofs pitched shallower than 6 in 12 should have a 4-in. lap. Very steep roofs, like mansards, can get away with a 2-in. lap.

Hilgert framed the roof of the new bathroom we added in the same manner as the house. Instead of 4x6 rafters, he used standard 2x8s; and for roof lath he used 4/4 by 2-in. white pine. Only along the ridge and the eaves do you have to sheathe the rafters. On our roof, Hilgert used wide 4/4 boards.

As a rule, you can get by with conventional framing if you're installing a standard slate roof like ours. But increasing rafter size by one nominal dimension (from 2x8 to 2x10, for example) would reduce the deflection of these members over the years, particularly with a snow load. We used standard ³⁄₁₆-in. slate, which weighs between 750 lb. and 850 lb. per square, depending on where it was quarried. If you're planning what's known as a textured or a graduated slate roof, you will need to beef up your framing considerably. These two roofing styles call for slate that's ⅜ in. to 2½ in. thick, which translates into loads of 1,500 to 6,000 lb. per square.

Though some slate roofers prefer to use conventional sheathing beneath a slate roof, we decided to stay with the original 4/4 lath, since most of it was in good shape. Hilgert also believes that an airspace directly underneath the slate allows it to dry out more thoroughly after a storm.

If you decide to install a slate roof over sheathing, the sheathing should be covered with overlapping layers of 30-lb. asphalt felt before the slate goes on. The felt protects the roof from weather while the slate is being laid, and also forms a cushion for the slates.

The rafters in the main part of our house were in excellent condition, and I knew that they could carry the weight of the new slate with no problem. After removing the old slate,

however, we found that some of the old hemlock roof lath would have to be replaced, along with a few of the wide boards at eaves and ridge. Some of this old hemlock had become so hard and brittle over the years that you couldn't drive a nail into it without causing entire runs of lath to vibrate. This, in turn, caused already installed slates to shake and pull free from their nails. So this old wood was replaced with new white pine.

**Cutting and hole punching**—The traditional tools for these tasks are a slater's hammer and stake (photo left). You probably won't find them at your local hardware store; I got mine at a flea market. New tools are available from John Stortz and Son, Inc., (210 Vine St., Philadelphia, Pa. 19106).

The hammer is made from drop-forged steel and has a leather handgrip. Between handgrip and head, the handle is flat, with one edge beveled sharp, so that the tool can be used to smooth rough edges of trimmed slate. Where the handle joins the head, there's a stubby pair of claws for pulling nails. The striking face of the head is small—about the size of a nickel—to minimize the risk of damaging the slate when nails are driven home. The other end of the head tapers to a fairly sharp point. This sharpened end is used to punch nail holes in the slate, and to perforate slate along a scribed cutting line. Once perforated, the slate can be broken, and the resulting jagged edge can be smoothed with the beveled edge of the handle.

The technique for punching and perforating takes time to master. It's a short, quick, well-aimed stroke that stops just after the hammer's metal point strikes stone. Smoothing a cut edge with the beveled handle is easier: just chisel the slate smooth. It's not a bad idea to practice your technique on a few broken pieces of slate before working on slate to be nailed up.

An alternate hole-punching method is to drive a nail through the back of the slate. Always work on the face that won't be exposed to the weather. This way, the slightly broken or beveled slate surface will face up.

Another important thing to remember is that there are right-handed and left-handed slater's hammers. You can tell the difference immediately if you hold the hammer in the wrong hand and try to trim a slate—the nail-pulling claw will get in the way.

The stake, a T-shaped piece of steel, supports slate when it's being trimmed. The short leg of the T comes to a point, so the stake can be driven into a rafter. For our roof, though, Hilgert drove the stake into a stump beneath a shade tree and did his cutting there. And to make simple, straight cuts, he often used a non-traditional tool that most slaters consider indispensable today—a tile cutter. The score-and-break technique used for straight tile cuts works fine for slate too.

**Nailing it up**—Roofing with slate doesn't differ fundamentally from roofing with other kinds of shingles. Overlap from one course to

the next should cover the nail holes of the lower course by at least 3 in., and the joints in one course should be staggered by at least 3 in. from those in adjacent courses. This means that you've got to cut some slates to keep the joints sufficiently staggered. If one course begins at the gable with a full (12-in. wide, in our case) slate, the next course will have to begin with a partial slate. Try not to use partial slates that are extremely narrow (3 in. or less), since these are especially prone to breakage.

Once the framing had been repaired, Hilgert's crew nailed a starter course of slate directly to the wide sheathing along the eaves, overhanging the framing by about 2 in. As shown in the drawing and photos on the facing page, this starter course is laid horizontally, with its length running parallel with the eave. It's best if the starter-course slates are installed face down. This way, the slightly beveled, chipped edge faces downward, creating a better drip edge. After the starter course, all slates should be installed vertically, with their beveled edges facing up.

Proper nailing technique is the most important part of applying a slate roof. If you're used to nailing wood or fiberglass shingles, you'll have to go easy when working with slate for the first time. You're pounding a nailhead that's surrounded by fairly delicate stone, and a single miss can ruin a good slate. A carpenter's hammer can be used, but the narrow head on a slater's hammer is less likely to break the slate surrounding the nail hole.

Nail holes are typically machine-punched at the quarry, but you'll have to hand-punch the slates that are used for hips, valleys and ridges. As shown in the drawings on the facing page, the nail head should sit just below the top surface of the slate. If it's driven too far, the slate around the hole will crack. If it's not driven far enough, the protruding nailhead will crack the slate that overlaps it.

The original slate on our roof had been nailed down with iron cut nails, most of which were in good condition when we removed the slate. Because of this, we decided to use galvanized roofing nails rather than copper nails. Copper is definitely the best choice for slate work because its longevity better matches that of the slate. But copper nails are also a lot more expensive (about $3.00/lb., compared with $.90/lb. for galvanized), so we're hoping that our hot-dipped galvanized nails will last as long as the iron cut nails did. If you want to use copper nails, the sources I've found for

***Framing.*** You can use solid sheathing beneath a slate roof, or roof lath, which was used on the roof shown here. Lath spacing is important, as shown in the drawing opposite, and the eaves and ridge require solid sheathing. Facing page, left: a rotten eave board in the old roof is replaced with new wood.

***First courses.*** The starter course is nailed horizontally to the eave sheathing (facing page, right); then the first vertical course follows. Adjacent slates should be butted together without overlapping. Vertical joints in successive courses should be staggered at least 3 in.

Illustrations: Peter Jennings

## Installing slate over roof lath

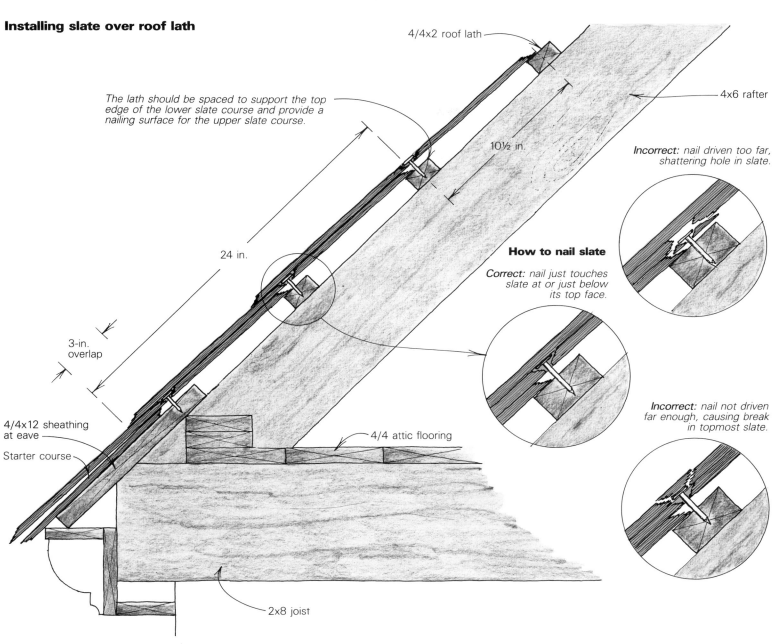

4/4x2 roof lath

4x6 rafter

The lath should be spaced to support the top edge of the lower slate course and provide a nailing surface for the upper slate course.

10½ in.

24 in.

3-in. overlap

4/4x12 sheathing at eave

Starter course

4/4 attic flooring

2x8 joist

**How to nail slate**

*Correct:* nail just touches slate at or just below its top face.

*Incorrect:* nail driven too far, shattering hole in slate.

*Incorrect:* nail not driven far enough, causing break in topmost slate.

them are Glendenin Bros, Inc. (4309 Erdman Ave., Baltimore, Md. 21213), Prudential Metal Supply Corp. (171 Milton St., East Dedham, Mass. 02026.) and Vermont Structural Slate (Box 98, Fair Haven, Vt. 05743). Correct nail length for standard ³⁄₁₆-in. slate is 1½ in., for either copper or galvanized nails.

As shown in the drawing on the previous page, each run of roof lath supports the top edge of the slate course below and also serves as the nailer for the following course. To line up successive courses, Hilgert's crew snapped a chalkline down the center of each strip of lath. The upper edge of the next slate course was then laid to this line, leaving about 1 in. of nailing space in the same piece of lath for the following course.

Until the first half-dozen courses of slate had been laid, the crew members could reach the work from the scaffolding under the eaves, or simply by standing on the attic floor. Reaching the higher sections of the roof was more difficult, because they couldn't walk on the installed slates without the risk of breaking them. To reach upper roof sections, they worked from a ladder that they built from 1x4s. The ladder rails are a pair of 1x4s positioned with their broad faces against the roof surface. This provides more even weight distribution than a conventional ladder. A 4x4, cleated across the top of the ladder, holds it in place against the ridges.

**Snow guards**—Snow and ice accumulation along the eaves can really damage a slate roof. The eave is often the coldest part of the roof, and snow that melts on warmer upper sections can slide down and refreeze at the roof edge. This added weight can cause eave slates to crack and break.

One way to prevent eave icing is to flash the eave with a continuous strip of metal, usually aluminum. The first slate course overlaps the top edge of the flashing by at least 3 in. Only a little snow or ice will stick to the metal before additional accumulations cause the icy mass to slide off and fall to the ground.

The more common approach to eave protection in our area is to use snow guards in above-eave areas of a slate roof. A snow guard (photos above and facing page) is a right-angled metal cleat that is nailed to the lath between slates. Its working edge sticks up above the slates, and is designed to hold snow in place on the roof, minimizing slide-down accumulations along the eaves.

Like our slate, the snow guards we used were recycled. We bought them from another local contractor who had salvaged them when he re-roofed a church in a nearby town. The cast-iron snow guards were at least 75 years old, very rusty, and spotted with roofing tar. We had them cleaned and hot-dip galvanized. All told, they cost us about $6.50 apiece.

On the main roof section shown in these photos, Hilgert installed the guards 4 ft. o. c. in the second and fourth courses. On smaller roofs, you could probably get by with only one row of snow guards. As shown in the photo on the facing page, one slate has to be notched to fit around the snow guard's installation strap. The following slate course then covers this strap.

**Flashing and ridge details**—As roofs go, the one shown here is simple—no hips, dormers or valleys to contend with, only a couple of chimneys. Because of this, installing the slate was fairly straightforward. But a more complex roof wouldn't be a problem for anyone who's familiar with the hip, valley and flashing details used with wood shingles (see *FHB* #9, pp. 46-50). Chimneys, dormers, skylights and sidewalls that penetrate or intersect with a slate roof should be step-flashed. Hilgert used copper flashing on our roof (with copper nails to avoid any problems with galvanic action), but aluminum, tin, lead and zinc have also been widely used.

Though closed and even round valleys are found on some slate roofs, the open valley is the most common. Install metal valley flashing for a slate roof just as you would for wood shingles. Standards set by the National Slate Association back in 1926 call for open flashing to be slightly wider at the bottom of the valley than at the top to handle the increasing vol-

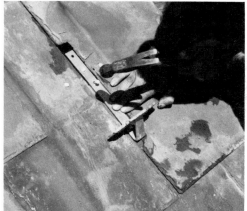

***Snow guards.*** Nailed to the roof lath between slates, these aluminum or cast-iron elements are designed to hold snow on upper sections of the roof, preventing damaging ice and snow accumulation at the eaves. One slate should be notched to fit around the mounting strap, as shown above. The following course covers most of the strap (facing page). Once you master the nailing technique, a roof with no hips or valleys can be slated quickly.

ume of runoff. Adding about ⅛ in. to valley slates in each succeeding course should create sufficient taper in the open valley.

There are also several options when you come to the ridge. The major ones are shown in the drawing at right. Hilgert finished off our roof with a strip saddle ridge. As the drawing shows, the final full course of slate on one side of the roof extends so that the upper edges of its slates are even with the solid sheathing at the ridge peak. Then these edges are overlapped by the final full course of slate on the opposite side of the roof. The final step is to nail down a second pair of overlapping courses, using partial slates that run lengthwise along the ridge. Nails in this last layer of "combing" slates are positioned so that they fall in the seams between the slates in the course below.

As with a wood shingle or shake roof, detailing at hips can get complicated. Saddle and flashed hips are popular, but you can also use a Boston hip (see *FHB* #12, p. 56).

Hilgert used no roofing cement to point the seams along the ridge or to cover the exposed nails in final combing courses. Like many slaters, he believes that cement isn't a requirement if a slate roof is installed properly. But in very rainy territory, or if you want to be doubly sure of your roof's weathertightness, use a high-quality silicone caulk to cover exposed nails and to point ridge-course seams. I had some doubts about leakage through the ridge, but the main roof of our house passed its first test for water-tightness the day after it was finished. A driving rainstorm, one of the last in an altogether too-wet spring, hit our part of Pennsylvania. Flashlight in hand, I went up to the attic to look for leaks. I could hear the rain pelting the roof, but not one drop of water found its way through.  □

*David Heim is a freelance writer based in New York City.*

## Ridge and hip details

**The strip saddle ridge** *relies on overlapping courses along the ridge for weather worthiness. The final courses on each side of the roof overlap along their top edges. They are then covered by two combing courses —slates that run lengthwise along the ridge, overlapping along their top edges and butting along their side edges.*

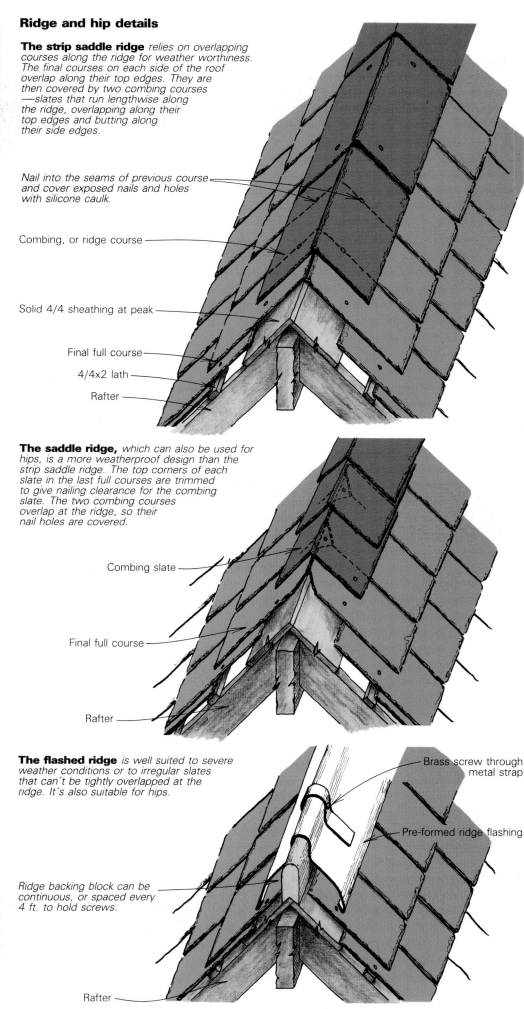

Nail into the seams of previous course and cover exposed nails and holes with silicone caulk.

Combing, or ridge course

Solid 4/4 sheathing at peak

Final full course

4/4x2 lath

Rafter

**The saddle ridge,** *which can also be used for hips, is a more weatherproof design than the strip saddle ridge. The top corners of each slate in the last full courses are trimmed to give nailing clearance for the combing slate. The two combing courses overlap at the ridge, so their nail holes are covered.*

Combing slate

Final full course

Rafter

**The flashed ridge** *is well suited to severe weather conditions or to irregular slates that can't be tightly overlapped at the ridge. It's also suitable for hips.*

Brass screw through metal strap

Pre-formed ridge flashing

Ridge backing block can be continuous, or spaced every 4 ft. to hold screws.

Rafter

# Putting on a Concrete-Tile Roof

## It takes nimble feet, a strong back
## and attention to flashing details to do it right

by J. Azevedo

**M**ost people with even a little construction experience would not hesitate to install an asphalt-shingle roof, yet they quake at the idea of putting on concrete tile. Maybe it's the mystery of flashing, where the undulating tile profile meets the straight line of ridge, or worse yet, the angled line of hip or valley. Maybe it's the uncertainty of laying a material that won't bend, not even a little, and is too thick to slip under a skylight flashing. Or, maybe it's simply the idea of moving all that weight around. After all, an average roof might need 16 tons of tiles, and that number carries with it the image of hauling coal, growing older daily and falling deeper in debt.

After presenting the case for choosing a tile roof in another article (see pp. 88-93), I started to wonder whether there was in fact a bona fide basis for the amateur's reluctance to roof with concrete tile. To find out whether it's a job best left to the professional, I went up on the roof with one: Marlen DeJong, principal in the modestly named A-1 Tile Company of Everson, Washington. After spending a week watching him install roofs on three houses over the course of a summer, I have concluded that tile roofing, though physically demanding, should not intimidate a conscientious builder. The interlocking tile is virtually foolproof. The only tricky spots—and these do require careful attention to detail—are the "joints" in a roof: the ridge, rake, valleys, hips and roof penetrations.

**Readying the roof—**A tile roof requires a little more preparation than a shingle roof. Special attention must be paid to setting fascia, applying felt underlayment and nailing on horizontal battens. DeJong asks the carpenters to run the eave fascia ⅝ in. past the rake and to set it 1½ in. above the plywood roof sheathing (top photo, facing page). Like a shingle starter course, the raised fascia lifts the first course to the same pitch as the other courses.

The felt underlayment forms a second line of defense should any water somehow sneak past the tile. DeJong uses 30-lb. roofing felt, 36 in. wide, lapped 3 in. at each course. At the eave, where the raised fascia would dam the water, he installs an "anti-ponding strip," which is a preformed sheet-metal flashing (available from most roofing-tile manufacturers) that takes the felt (and any runoff) smoothly over the raised fascia. In areas that get snow, Dejong doubles the felts along the eaves and seals their edges with roofing tar to seal out any water that might get beneath the tiles as a result of ice damming.

At the ridge, DeJong normally takes the felt over the peak and down the other side a foot or so. If there are ridge vents, however, he holds both the sheathing and the felt back 2 inches from the peak. At valleys, he runs the felt from one slope across the valley and a few feet up the other side and cuts the opposite felt to lie right down the center of the valley (variations include interlacing the felts, and even adding another strip vertically down the valley). DeJong lays standard valley flashing on top of the felt (middle photo, facing page). The flashing is 24 in. wide and shaped like a "W" in profile, with the long edges turned up and inward to contain runoff. The flashings are secured with 8d nails bent over their long edges. Where valley flashings overlap, the bottom piece can be nailed near its top edge. The next piece overlaps it by 6 in., and the joint is sealed with roofing cement. Where the valley flashing meets the anti-ponding strip, he crimps the valley to match the change in pitch.

Felts and flashings are the weak spots in a roof that is supposed to last better than 50 years. DeJong offers his customers upgrades— copper or lead flashings instead of painted galvanized steel; double felts or 45-lb. felt instead of 30 lb.—but almost no one goes for it, an attitude that seems shortsighted.

**Battens—**With the roof felted, DeJong starts to lay out the battens—the horizontal strapping (typically 1x4s or 1x6s) that will support the tile. The battens define the course spacing, so the layout must be precise (bottom photo, facing page). First, he staples ¼-in. by 2-in. lath over each rafter. The lath raises the battens, allowing any moisture to drain down to the gutters. Then he sets the first batten parallel to the eave, with its uphill edge 14½ in. back from the front edge of the eave fascia. He goes then to the ridge and sets a batten 1¾ in. down from the peak. Next comes a calculation to determine the course spacing. He divides the distance from the ridge batten to the eave batten by 13½ in. (the ideal course). If the result is not a whole number (and it seldom is), he takes the next highest number of courses and divides the distance by that number to get the course spacing. If you don't like math, don't worry. Tile manufacturers provide tables that give course spacing for any roof. Once DeJong has the spacing, he marks out the courses on the felt, snaps chalklines, and nails down the remaining battens, staggering the joints between them.

The only complication in laying the battens is when two roof slopes of unequal length meet as one at the ridge, such as at staggered eaves. In that case, the course spacing for one roof must be used on the other side as well. That usually results in a short course at the eave.

**Loading the roof—**With a roof that averages 16 tons, three things are certain: you don't want to carry the tiles up a ladder; you don't want to stack them all in one place; and you don't want to move a tile very far once you put it down. Fortunately, roofers have worked out a system for avoiding the three "don'ts."

Tiles arrive on pallets, and suppliers will either deliver them to the roof or arrange delivery at extra cost. From there it is up to the roofer to unload and distribute the tiles.

DeJong guides a pallet, swinging from a boom truck, to a spot near the ridge and has it lowered until the front just rests on the roof. He dons a pair of heavy gloves and unloads a bundle of six tiles near the peak. He continues this way along the ridge, spacing the bundles 2 ft. o. c. Then he moves down three battens and lays another row of bundles. He works his way down the roof; the last row of bundles rests on the third batten up from the eaves. Any extra tiles go back up near the ridge or near hips.

Before stacking the roof, DeJong looks at the batch dates on the pallet wrappers. Each batch may vary slightly in shade, so he keeps all tile from the same batch on one section of roof. Other roofers like to distribute tiles to mix the variations into the roof.

**Laying the field—**The large interlocking field tiles go down without much trouble and cover a roof quickly. DeJong starts from his left as he lays the first eave tile (bottom photo, facing page), hooking the lug on the bottom of the tile over the first batten and holding it back from the rake about 1 in. to 2 in. The next tile interlocks with the first along their common edge (called the "waterlock"). As he continues along the eave, DeJong checks the waterlock on each tile to make sure it's clean.

From *Fine Homebuilding* (February 1991) 65:56-61

Sometimes a bit of the slurry color coating on the tile has clogged the waterlock, and he must scrape this out or the tiles will not fit neatly. He lays them loosely, with a gap of $\frac{1}{32}$ in. to $\frac{1}{16}$ in. to allow for structural movement. When he gets to the opposite rake, he hopes to end the course with either a whole tile or a half tile. If this doesn't happen, he can usually adjust the course, squeezing the tiles or pulling them apart slightly to fit. The goal is to have one of the three tile "barrels" (the ribs or humps of a tile) at each end of the course to direct water away from the rake, toward the watercourses on the field tiles. When he is satisfied with the fit, he drops an 8d galvanized nail through the hole in each tile and taps it snug. How snug? "If you break the tile, you know it's too tight." Yet, you don't want to leave it sticking up. If you do, it will pop the corner off the next tile up the first time you walk on it.

Once the first course is nailed, DeJong measures in from the rake to the second joint and makes a mark. He marks other joints about every 10 feet along the course, and then snaps parallel lines all the way to the ridge. These marks will serve as guides to keep the courses aligned vertically up the roof.

DeJong lays the second course much the same as the first, only he starts with a half tile instead of a whole one (to learn how to cut and break tiles, see the sidebar on p. 87). He continues this way, course by course up the roof, staggering the joints and lining them up to the chalk marks, thumping each tile with the rubber handle of his hammer and listening for the telltale rattle of a tile that is not lying flat against its neighbors or the batten. As he falls into the rhythm of laying field tile, he seems to deal out these 10-lb. tiles with all the ease of a casino blackjack dealer.

On roofs with less than a 12-in-12 pitch, he nails only every other course. Above that, or if called for by the manufacturer's specifications or local codes, he nails every tile. On really steep pitches or in high-wind areas, he also fastens down the nose of each tile with a copper hurricane clip (available through the tile supplier).

**Hips and valleys**—Hips and valleys are basically mirror images and the tiling is much the same. It's a matter of getting the angles right and holding the cut pieces in place.

To tile a hip, DeJong starts by toenailing a 2x4 nailer on edge along the hip, holding the lower end back about 4 in. to 5 in. from the eave. Then he lays field tile as close as he can to the nailer without cutting any tiles. Near the nailer, he can use broken tiles that he has been saving. After DeJong runs a circular saw along a chalkline snapped parallel to the hip (bottom left photo, next page), he blows off the dust and tosses the scrap. The cut pieces should fit tightly against the nailer (top photo, next page) once he inserts two standard field tiles into each course.

DeJong fastens the cut pieces two different ways, depending on their size. For large pieces,

At the eave, a raised fascia supports a sheet-metal "anti-ponding strip." The strip acts as a starter course to raise the level of the first course of tile into plane with the rest of the roof.

Valley flashings are 2 ft. wide, with an angled lip along both edges that directs runoff back toward the flow line. DeJong secures the flashings to the roof with 8d galvanized nails driven outside the flashing edges; he then bends them over the lip.

Before the battens go down, DeJong staples lath to the deck over each rafter. This allows any moisture that may sneak under the tiles to escape easily. Here DeJong begins the first course of tile, hanging tile lugs on the first batten.

which still have part of the lug, he hooks the tile over the batten, drills a hole near the top, and nails it in place. For small pieces, he drives a nail partway into the deck near the nailer as a platform to support the edge of the tile. He hooks the bottom edge with another nail driven sideways into the nailer.

If a small angled tile finishes off the eave course where flashing prevents him from driving nails, he cements the piece to the next whole tile with sticky black roof tile cement (Fields Corp., 2240 Taylor Way, Tacoma, Wash. 98421; 206-627-4098) in the top 2 in. of the waterlock. Once the tiles are set, he then tapes the two tiles together to hold them until the cement sets up. DeJong flashes the joint by molding 4-in. flexible flashing tape (Aluma-Grip-701, Hardcast Inc., Box 1239, Wylie, Texas 75098; 214-442-6545) down from the top of the nailer and onto the tile.

Finally, he caps the hip with special trim tiles (small photo, below left). He test-fits a hip tile, like putting a saddle on the nailer, and checks that the bottom edges just rest on the barrels of the adjacent field tiles. If the nailer is too low, he shims it with lath. Then he starts the hip with a bullnose end cap nailed to the nailer. He spreads tile cement over the nail, drops a standard hip tile down on the nailer, slides it down into the cement until the bottom lug catches on the tile below and nails it in place. He continues up the hip to the apex, where he either miters the hips to meet the ridge (a ridge tile will later slip under this mitered joint) or caps it with a special apex tile. In either case, he flashes under the apex with flexible flashing or lead.

Some roofers prefer to bed the hip tiles in mortar. If you choose to do this, be sure not to pack the joint. Just tuck mortar under the edges. Otherwise, the mortar will wick water up into the hip.

Like hips, you can cut and fit valley tiles one at a time or cut the whole row at once. But when it comes to securing cut tiles to the roof, DeJong alters his technique. For large pieces, he strings a length of 14-ga. galvanized wire through the nail hole and ties the wire back to a nail driven just outside the valley (top photo, facing page). In this way he avoids punching holes in the valley flashing. If any of the cut tiles tips or rocks, he sticks a chunk of broken tile underneath to support it (some roofers tuck mortar under the cut edge).

**Rakes and ridges**—Rakes are trimmed with special tiles that are nailed to the rake fascia. Rake tiles resemble hip tiles but are formed to a tighter "V." They are interchangeable for right and left rakes.

Balancing precariously on the eave corner, DeJong sets the first rake tile on the end of the eave course and slides it up against the nose of the next field tile up. He holds the side of the rake tile plumb and carefully drives two nails through the prepunched holes. Working his way up the rake, he butts a rake tile against each course of field tiles until he gets within

*Hip trim.* **It's most efficient to gang-cut angled tiles for hip or valley intersections. Note in the foreground (left photo) the position of the nail holding the chalkline. The distance from it to the corner of the tile to its right is the same as the distance from the tile nearest the hip to the edge of the nailer. Trimmed hip tiles tuck against the 2x4 nailer (top photo). Short pieces are supported at their lower edges with a nail driven horizontally into the 2x4. Asphalt roofing cement, along with flashing tape, can also be used to glue smaller tiles to their neighbors. Hips are finished with trim tiles affixed to the 2x4 nailer with 8d galvanized nails topped with a dab of asphalt cement (photo above). The shim under the second trim tile ensures a sturdy base for the tile.**

***No holes in the valleys.*** At valleys, tiles are secured without penetrating the flashing. On the near side, a tile is wired to a batten. On the far side, DeJong tapes a piece of tile to its neighbor. Asphalt cement between the short piece and the overlapping tile will further secure it.

Rake tiles that are V-shaped in section trim the gable ends of roofs. At the top, they are mitered and sealed with asphalt. Trim tiles similar to the hip tiles finish off the ridge.

DeJong's preferred method to seal chimneys and sidewalls against the weather is with lead step flashings shaped to fit the irregular contours of the tile.

Another way to tile up a sidewall, chimney or skylight is with a sidewall flashing. Tile edges overlap the raised lip of the sidewall flashing, spilling runoff into the flashing channel.

A vent is sealed by folding the site-cut tabs of a lead jack over the lip of the vent.

tangle of 2½-lb. lead sheeting and bends it in half lengthwise. He spreads tile cement on the bottom half and presses the flashing into the corner, with the bottom edge even with the headwall flashing. The lead can be molded to conform to the shape of the tile below.

DeJong prefers the step-flashing system for sealing along chimneys or sidewalls (bottom center photo, previous page). He lays the next tile course right up to the side of the chimney and sets another bent, gooped, 12-in. by 16-in. piece of lead step flashing atop that tile, with the nose of the flashing even with the nose of the tile.

Once he works flashing up to the top of a skylight or a chimney, DeJong fits in a piece of rigid "back flashing"—a sheet-steel "V" that extends up the back curb of the skylight and 16 in. up the roof. At the sides, the back flashing extends 6 in. past the edge. DeJong blocks under the flashing with battens and tile pieces to raise it up just high enough that it sets on top of the step flashing. To make the joint, he cuts down the crease of the last step flashing almost to the corner of the curb and folds the vertical flap around the corner and against the back curb. He also cuts down the top flap of the back flashing to the corner and bends it around the corner and against the side. When the fit is right, he pulls the back flashing, goops all the overlapping flaps and corners with roof tile cement and presses the back flashing back down into place. The next course of tiles runs across the top, sitting on the back flashing.

If the skylight or chimney is wider than 4 ft., DeJong hires a carpenter to frame a cricket. Then he tiles around it as he would a valley.

The other way to flash the side is with a single piece of sidewall flashing (sometimes called J-flashing). This is a sheet-metal pan turned at a right angle to go up the wall. It has a raised lip along the outer edge of the pan like valley flashing. The flashing rides on battens, and the edges of tiles drop into this pan, which channels water down the side and spills it onto the course below the chimney or skylight.

At least one skylight company (Velux-America Inc., P. O. Box 3268, Greenwood, S. C. 29648; 800-888-3589) offers its own tile-compatible flashing kit using the sidewall flashing approach (bottom right photo, previous page). If you buy these skylights, you should probably consider getting the flashing as well. It's not hard to install, despite the instructions.

The flashing at the base of the Velux skylight shown here is made of ribbed lead sheeting. Before one can be installed, the protruding barrels of the tiles need to be lopped off so the sheet will lie flat. Also, the noses of the side tiles are notched to clear the lip of the sidewall flashing, and the tiles are tied back with wire to avoid puncturing it.

**Tiling around a vent pipe**—Flashing around a plumbing vent is quite a bit easier than sealing larger openings. The secret is the malleable lead jack (photo above).

DeJong tries to coordinate his work with the

one tile of the ridge. Ridge and rake will meet at the apex, so he first must prepare the ridge.

Along the ridge he has already toenailed a vertical 2x3 nailer, stopping it back 6 in. from the gable end. The top course of field tile nearly butts this nailer. Now he molds flexible flashing tape down the nailer and onto the tiles, as at the hips, to seal the ridge joint. At this point he can cut the apex of rake and ridge.

DeJong miters the opposing rake tiles at the peak so that they meet with a neat, tight joint. He drills another nail hole in each trim piece, sets nails in the rake as spacers to hold the pieces plumb and square to the rake, and nails the mitered trim with two nails each.

To cap the ridge, DeJong butts a ridge tile square against the mitered rakes (bottom left photo, previous page) and nails it to the 2x3 nailer, as at the hips. Another way to do this detail is to notch the bottom of the ridge tile so that it overlaps the mitered rake joint. He goops the mitered seam and stuffs more tile cement into the arch of the ridge tile where it meets the rake. At first the shiny black cement

looks unsightly and out of place, but it will eventually fade and the color then will blend with the tile.

DeJong completes the ridge by lapping trim tile, nailing and cementing the joints as he did on the hips. When he gets close to the opposite end, he checks the distance remaining and adjusts the spacing to get the last ridge tile to meet the apex of the rake. Or, he simply cuts the last tile to fit.

**Chimneys and skylights**—DeJong tiles right up to the base of a chimney or a skylight as if it were a ridge. If it works out that this last course is short, he cuts the tiles and uses the bottom portions, drilling them for nails and shimming them as necessary to maintain the roof pitch. Then he seals the joint with Aluma-Grip tape, and caps that with rigid counter-flashing (sometimes called "headwall flashing"), which extends down 6 in. over the barrels of the tiles below. Incidentally, he uses this same detailing where a roof meets a headwall.

To flash a bottom corner of a boxed-in chimney, DeJong cuts a 12-in. by 16-in. rec-

plumber. Ideally, he would like the vents to come up between the battens, in the center of a tile. This location produces the neatest, most water-tight job with the least fuss. The vents should project about 15 in. above the deck.

As he tiles up to the vent, DeJong measures, marks, and cuts the tile to fit around it, and then sets the tile in place. He slides a lead roof jack over the vent and onto the top of the tile, square to the roof. With the rubber handle of his hammer, he dresses the jack to conform to the curves of the tile, and he trims the back of the flashing even with the back of the tile. He slides the jack up, goops roof-tile cement around the hole in the tile, and slides the jack back down, bedding it in the cement. Then he folds inward site-cut tabs in the top of the jack to seal the vent.

**Safety**—To help you keep your footing, DeJong recommends you wear rubber-soled shoes (athletic sneakers or deck shoes). Concrete dust is slick; blow off the roof after you cut tiles. And, stay off a wet roof.

On steep pitches, the situation gets more precarious, and safety becomes more critical. For modestly steep pitches, say 8-in-12 or 12-in-12, stay off the tiled portion, walk on the battens and make "ladders" to get across tiled slopes. Steeper than that, plan to hang on the batten with one hand, and consider a scaffold or safety net at the eave to give you a second chance. Or, maybe this is the time to call a professional. The National Tile Roofing Manufacturers Association (NTRMA, 3127 Los Feliz Boulevard, Los Angeles, Calif. 90039; 800-248-8453) makes available a list of tile manufacturers, who can in turn recommend competent installers. □

*J. Azevedo overcame his acrophobia to research this article. Photos by the author.*

## Cutting and breaking concrete tile

Concrete tiles are only sand and cement; you can cut them with a carborundum wheel mounted on your circular saw. Still, abrasive masonry wheels are exasperatingly slow, and you will wear out several on a tile roof. If your roof is cut up by a lot of hips, valleys, and skylights, DeJong recommends that you consider buying a diamond blade. He cuts tile with a 12-in. diamond blade mounted on a portable electric cut-off saw (for more on these tools, see *FHB #62*, pp. 80-84). The diamond chews through tile about as fast as a circular-saw blade cross-cuts 2x fir.

Breaking tiles is easy, but getting them to break where you want takes a little more finesse. DeJong generally breaks, rather than cuts, half tiles for starting courses at the rake. First he cuts or chips off the lugs and reinforcing ribs down the center of the underside of the tile. Then he cradles the tile, face side up, in his arms like a baby and raps the center with a hammer. If all goes well, and it generally does, the tile splits in half down the center (photo below right).

To drill a small hole in a tile, DeJong uses a standard masonry bit. For larger holes, as for vent pipes, he scores a square with his diamond blade and knocks out the opening with his hammer.   —*J. A.*

After laying out a chalkline parallel to the hip, DeJong uses a 12-in. dia. blade to make clean, fast cuts in the concrete tile.

To divide a tile without a saw, remove the lugs from the center of the tile's underbelly, flip it over and give it a rap in the middle with a hammer.

# Tile Roofing

## Clay or concrete, a tile roof will outlast the next generation

by J. Azevedo

In an 1895 technical book on tilemaking, author Charles Davis listed as a disadvantage of tile roofing that it is "anything but handsome." He was not the first to make disparaging remarks about the appearance of tile. Its popularity has cyclically risen and fallen in this country for the past 300 years. The recent resurgence of tile in the residential American roofscape—along with the consequent assumption that many homeowners must find tile to be quite handsome after all—reminds us that tastes do change. And in the case of tile's return to favor, changing taste coincides with good sense—the advantages of tile roofing far outweigh its disadvantages.

All tiles worth considering come with a 50-year guarantee, a measure of their durability in all climates. In wet climates, tile will not rot. Moss and fungus cannot get a foothold, bugs find it unappetizing, and rodents and squirrels chip their teeth on it. In cold climates, although tile will not shed snow as a metal roof will, well-made tile will withstand repeated freezing and thawing (most Swedish houses are roofed with tile). In coastal areas or cities, tiles can be used with impunity because they are unaffected by salt spray or pollution. In hot, dry areas, tiles will not split or crack. Also, tile roofing is available in light colors to reflect the sun's rays. A tile roof will not stop a fire that starts in your kitchen, but it will keep a fire in your neighbor's kitchen from spreading to your house through his roof. With its Class A fire rating, tile has the approval of every fire code from Boston to Los Angeles.

Finally, tile roofing in its many forms has been around so many years that its use in a new house can recall a certain region or historical period. Some tile profiles are synonymous with California missions or French country estates or oriental temples. Yet, not every tile has a singular connotation. A modern concrete tile with a profile loosely based, for example, on a Mediterranean tradition can look quite at home on a house in the Northwest. In fact, the array of tile profiles and colors available today give tile the virtue of versatility. Tile roofing can ripple lightly across the roof or sit elegantly in solid silence; it can form neat symmetrical vertical ribs or take playful dips and leaps. If only Charles Davis could see them now.

**It began with clay**—Clay-tile roofing originated in neolithic China and later, independently, in the Middle East. From these two loci, tile roofing spread throughout the Orient and Europe. In China, tile roofing has become vernacular. The Japanese adopted tile roofing, adapted it to their harsh climate, and now

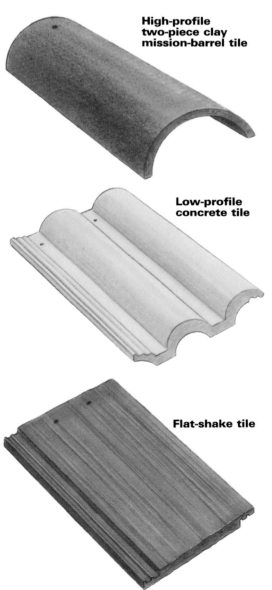

**High-profile two-piece clay mission-barrel tile**

**Low-profile concrete tile**

**Flat-shake tile**

have a major tile industry, including the largest clay-tile factory in the world. The ancient Greeks and Romans roofed their buildings with tiles of clay or, to show off, marble, and the Europeans picked up the clay-tile idiom and carried it to the present.

European settlers brought the tradition of tile roofing to America, starting at both coasts and working towards the middle. On the East Coast, Dutch settlers first imported tiles from their homeland but by 1650 were making them here. In the West, Spanish missionaries brought tile roofing along with their religion.

In California, tile folklore recalls how Indians made the first mission tiles from riverbank clay, formed across the curve of their thighs. Modern manufacturing methods, less sensual but more precisely controlled, generally start with shale, which is crushed to a fine powdery clay. The clay is mixed with water and kneaded (pugged) to the consistency of cookie dough.

To form simple clay tiles, moist, plastic clay is extruded through a die, like dough through a cookie press, and sliced into lengths. More complicated shapes come from pressing the clay into molds. Some ornamental tiles are even sculpted by hand. The formed tiles dry in a room kept at the temperature of a warm summer afternoon, and from there they go to a kiln for firing. For their trip through the kiln, the tiles are either stacked on a refractory (like a small railroad car) that is pulled through a tunnel kiln, or laid on ceramic rollers that convey individual tiles through a roller-hearth kiln. Regardless of the method, the object is to raise the temperature of the clay to the point of vitrification (about 2000° F), at which point the clay minerals lose their individual identity and fuse.

Clay tiles are available in colors to suit any taste, from subtle to outlandish. Right out of the kiln, the tile has the natural earth-tone of the original clay. A nearly pure clay will fire to buff, while iron-oxide impurities impart the terra-cotta color typically associated with Spanish tiles. Manufacturers can also mix clays to get a range of natural colors.

To make variegated tiles, a manufacturer can inject a burst of gas into the kiln, giving the tile a random scorched streak called "flash." Another way to color tiles is to spray a

High-profile tiles are typically made of clay, in either the traditional two-piece mission-barrel style, or the one-piece "S" tile shown here. To discourage nesting, this roof has clay plugs, known as "birdstops," under the courses of tile adjacent to the eaves.

Low-profile tiles are designed to interlock at the edges to keep out the weather. These variegated concrete tiles have been colored with a slurry of cement and iron-oxide pigments.

Flat tiles can be colored and textured to resemble traditional roofing materials such as slates, or the wooden shakes suggested by the dark, striated patterns of the concrete tiles on this roof.

thin creamy layer of clay (called a slip) onto the tile before firing. The tile will then take on the color of the slip. The most dramatic, and most expensive, way to color tile is with a glaze (photos, p. 93). The metallic pigments in the glaze, when fired, melt to a glossy, vitreous, richly colored surface, much like that found on ceramic tile used indoors. Manufacturers who offer glazed roof tiles stock a range of deep brilliant colors and can also brew custom colors.

The long history of clay-tile roofing has produced a variety of styles (drawing, p. 88). The industry typically divides them into three groups: high profile, low profile, and flat. Today you'll find examples made of clay or concrete in all three categories.

Currently, the most popular style is the high-profile "mission" (top photo, previous page). It comes in either the traditional "two-piece barrel" (the concave trough—or *tegula*—and the convex cap—*imbrex*—which is a trough tile turned upside down) or the newer "one-piece S," which joins the trough and cap into a single tile.

Low-profile tiles have a more precise, interlocking look (photo below). Most low-profile tiles are made of concrete, and are characterized by linear bevels, arches and grooves. They are made to interlock along their edges (drawing, facing page) to keep out wind-driven rain and snow.

Flat (or shingle) tiles are just that. They range from simple unglazed rectangles of clay or concrete to textured and glazed flat tiles meant to resemble the variable topography of slates or the coarse pinstripes of thick wooden shakes (bottom right photo, previous page). Like the low-profile tiles, flat ones typically have interlocking edges to keep out the weather.

A few manufacturers can do custom work to match an existing tile that has gone out of production, or to produce decorative finials or other roof ornaments. Ludowici-Celadon, Inc.

(P. O. Box 69, New Lexington, Ohio, 43764) and Vande Hey Raleigh (1665 Bohm Dr., Little Chute, Wis. 54140) are two such companies.

**Concrete-tile roofing**—Compared with clay, concrete tile is the new kid, with a history of just over 150 years. A German farmer is credited with making the first concrete tiles in the early 1800s for his barn. Commercial production started in Bavaria in the mid-19th century. When concrete-tile roofing spread to the rest of Europe around the turn of the century, manufacturers began to add color to the tiles. Still, acceptance of the new product was slow to come. A 1930's book on tiling noted that concrete tile, "…generally considered a new industry, is subject to a good deal of prejudice," which the industry countered with a vigorous public relations campaign. One company even named itself "The Everlasting Tile Co." Europeans no longer need to be convinced—90% of European houses are roofed with concrete tile. The same holds true in Australia.

In the U. S., concrete tile has been around since the early part of this century, but the industry languished until recently. In the early 60s, the Australians first brought their enthusiasm for concrete tile to California, along with machines for making high-quality extruded-concrete tiles. Since then, concrete tiles have become firmly established in the roofing market. Now more than 30 major plants and many smaller ones sell concrete tile throughout the U. S. and Canada.

The typical concrete tile (drawing facing page) measures 13 in. wide, 16½ in. long and about ⅝ in. thick. It weighs about 10 lb. Along both edges runs an interlocking channel, and across the front, underneath, are two or more transverse baffles that rest on the tile below and block blown rain and snow. At the back, projecting lugs allow the tile to be hung on roof battens. The surface texture is either orange-peel smooth or, for tiles that are meant

to simulate wood shakes, striated. (For a look at how these tiles are made, see the sidebar on the facing page.)

Color was added to concrete tiles originally to mimic the terra-cotta of clay tiles, the grey-browns of wood shakes, or the grey-greens of slate. Most people still prefer these earth-tones, yet tiles are also available in striking blues, greens and even white.

Concrete tiles are colored by one of two methods. The first is to add iron-oxide pigment to the batch mix, which produces a uniform color all the way through the tile. A less expensive method is to coat the tile with a slurry of cement and iron-oxide pigment. Use of this technique also allows the manufacturer to add highlights of a second color, creating a shaded or variegated effect.

Regardless of the coloring method, tiles are additionally sprayed with a clear acrylic sealer. The sealer not only helps the tiles cure properly but also makes sure that any efflorescence (the white powder—free lime—that surfaces as concrete ages) is driven out the underside of the tile rather than out the top where it would spoil the appearance. As a side-effect, the sealer gives the tile a slight gloss. However, the gloss wears off in a few seasons, and the color softens to its true matte finish. Before or after the tiles are installed, they can be painted with standard acrylic paint to change their color, much like painting a stucco house. Painted tiles require periodic rejuvenation.

**Installing the tile roof**—Because tile lasts so long, all other components of the roof should be designed to give equally long service. Tile manufacturers all specify these details in their brochures. Some list minimum standards and secretly hope their customers will use better materials if they can afford them. Other manufacturers, who have seen their own tiles outlive the flashings and felts, now specify only the higher-grade materials—copper flashings, heavy felts (coated base sheets of more than 30 lb., built-up membranes or single-ply roof membranes), and copper nails. For slopes that are less than 3-in-12, tile roofing should be considered decorative only. The real roof is installed underneath.

The procedure for installing a tile roof varies somewhat with the tile pattern, and each manufacturer covers the specifics in their installation instructions. Basically, the roof is first felted, and horizontal and vertical guide lines are chalked to indicate the courses. Layout is critical because any deviations stand out against the pronounced vertical pattern of most tile roofing. If the tile is designed with lugs to hang on battens, those battens are nailed on now (some manufacturers approve of hanging their tiles on spaced sheathing over heavy felt underlayment draped over the rafters). Then the tiles are loaded onto the roof so that they are evenly distributed and within easy reach.

The first course rests on either a raised fascia, a cant strip, special under-eave tiles or, in

***Low-profile concrete tiles.*** **After the roof is felted, horizontal battens are affixed to the roof sheathing. Lugs at the top of the tiles bear against the edge of the battens to hold them in place. In seismic zones and windy areas, the tiles should also be nailed down.**

## Making concrete tile

Back in the 1920s, commercial concrete tiles were made by hand. Workers would pour a "moderately stiff paste" of mortar into wooden or metal molds and then strike off the surface with a profiled screed. The green tiles cured in the molds for a week or more before they were separated and shipped. Today the process is automatic. To see the process firsthand, I visited the SpecTile (now Monier) tile plant in Salem, Oregon.

As I watched the blue Skandia tile extruder whir and thunk, then general manager Rick Olson explained to me what was going on. A conveyor belt fed concrete into the hopper of the extruder. The concrete was not the slurpy mix I have poured for walks and walls but a rich, dry mix (5% to 8% water; 20% to 25% cement) that felt more like castle-building beach sand. The dry mix makes a stronger tile that cures more quickly. However, such a dry mix cannot be worked by hand to the required density. That takes high pressure from the extruder.

The extruder plopped a measured amount of concrete mix onto a steel tile mold, called a "pallet," formed to produce the complex profile of interlocking ridges, baffles and lugs on the underside of a tile. The top of the tile was rolled roughly to shape. Then, a cam grabbed the bottom of the pallet and forced it under tremendous pressure beneath a forming die that shaped and troweled the top profile. As the continuous mat of pallets emerged from the extruder, a guillotine knife sliced between the pallets to separate them into individual tiles (photo right) while another arm

punched holes for the nails. Their Swedish-designed, automatic tile-making system had been rated at one tile per second, but Rick told me they were pushing the equipment to 72 tiles per minute and that the extruder might be capable of cranking them out even a little faster.

Rick stopped the machine and pulled a newly formed tile off the tie line. "Here's one way to tell a quality tile," he said, pointing to the interlocking edge. The baffles and interlocking edges were crisp, knifelike edges.

The line of palleted tiles, resembling a nearly gridlocked freeway at quitting time, slid along tracks to the coating machine.

Photo: J. Azevedo

There, a pigmented cement batter drizzled onto a spinning horizontal brush, which flicked the slurry onto the tiles passing underneath. The dry-mix tiles sucked up the slurry, bonding the color to the surface. Another coating machine added a highlight color, or "flash," to the tiles. Finally, the tiles were sprayed with an acrylic sealer.

At this point the tiles had their final shape and color, but they still needed to be cured. We walked around back of the line to the insulated curing chamber, a space perhaps the size of an average house, warmed by the heat of curing concrete. Normally tiles spend 12 hours in this womb before being pushed out once more into the world.

Out of the curing chamber, the tiles rode the conveyor to the "depalletizer," where the pallets were separated from the tiles and returned to the extruder for another cycle. The cured tiles continued down the conveyor to where they were inspected for defects, stacked on recyclable wooden shipping pallets, shrink-wrapped, and moved out to the yard with a fork-lift. Olson, obviously proud of the company's quality control, told me that they must reject only one half of 1% in the plant, most of those for color problems.

As we walked around the yard, among white plastic cubes of bundled tiles, I asked Rick about business. "Well, these are all sold. I guess we could make more, but our employees like to spend nights and weekends with their families," he joked. And what about problems? "Our biggest problem is finding enough qualified installers who are willing to do a quality job." —*J. A.*

---

the case of mission-barrel tiles, a "birdstop" that also plugs the curved voids under the eave course (top photo, p. 89). Once the eave course is laid loose, checked for fit, and then nailed, successive courses are lapped over the one below as the roofer works diagonally up the roof.

Before the turn of the century, tiles on European roofs were fastened with two oak pins. Nowadays, nails will do just fine. The nails should be corrosion resistant at least; copper is better, especially in areas near the coast or affected by pollution. With standard concrete tiles on low-pitched roofs (less than 5-in-12) with battens, the field tiles need not be nailed. Their weight alone will keep them in place. The nailing schedule increases with pitch. In areas with high winds, all tiles are nailed and their butts secured with metal clips (available through your tile supplier).

With mission-barrel tiles, the fastening is a little more complicated (drawing next page). The trough tiles rest on the sheathing, so they can be nailed, but the cap tiles ride on the trough tiles, well above the sheathing. These tiles are generally fastened with copper wire that is either nailed back to the sheathing or tied to vertical copper strips running below

*The anchor lug at the top of the tile rests on furring strips or sheathing, while the baffles at the bottom of the tile match and engage the lower course of tiles, preventing rain or snow penetration. The interlocking channels intercept rain blown from the side and direct it down the roof.*

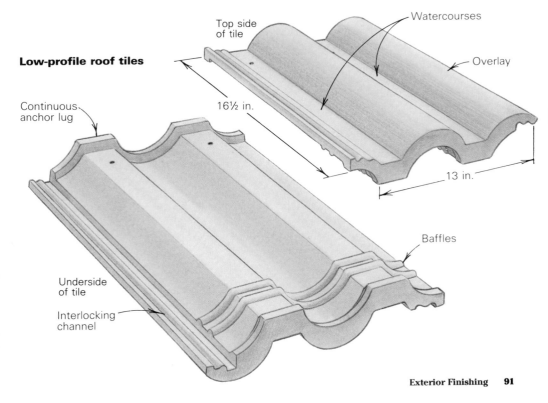

Low-profile roof tiles

Top side of tile

Watercourses

Overlay

Continuous anchor lug

16½ in.

13 in.

Baffles

Underside of tile

Interlocking channel

the tiles. Another method is to use tile nails, some of which look like gate hooks. Still another option is to nail vertical battens to the roof and nail the caps to these raised strips.

At the ridge or hip, concrete tiles are butted against a raised stringer. The joint is sealed with a flexible flashing, such as lead, and capped with ridge trim tiles affixed to the stringer. In place of flashing, some roofers prefer to point mortar under the edges of the ridge tiles to close off the watercourses. Care must be taken not to pack the joint fully because the mortar would then wick water up into the ridge. Clay tiles may be detailed the same way although some manufacturers offer special tile fittings to close the ridge and hip. With mission-barrel tiles, ridge and hip caps are usually mortared down. In heavy snow areas a tile roof, like any other kind, needs either extra felt or flashing that extends 2 ft. above the line of the exterior wall, or a ventilated roof deck (sometimes called a cold roof) to prevent damage from ice damming.

Once a tile roof is laid, it should need little or no maintenance. In fact, it is best to stay off of it. While tiles are strong enough to support careful foot traffic, someone clomping around on the roof risks breaking a tile. If you must walk on a tile roof, pick your way carefully across the reinforced or supported section of each tile, toward the nose (over the lugs) of concrete tiles and in the troughs (or "pans") of clay tiles. Mortared hips and ridges also

make good paths. Consider laying down a sheet of plywood to distribute your weight, or fill burlap sacks with sawdust and walk on these "pillows."

**The question of weight**—Try this experiment. Go up to one of your friends and say, "I'm thinking about putting on a tile roof." Chances are your friend will reply, "But aren't they heavy?"

The answer to that question is, of course, "Certainly." A typical tile roof weighs about 900 lb. per square—at least three times as much as wood or composition shingles. In the past, tile's reputation for being heavy has discouraged its use, but a heavy roof is not necessarily a bad roof. In the case of new houses, most are designed to carry three composition roofs: the original plus two re-roofs at intervals of 15 years. A tile roof weighs about the same. For new construction, then, a tile roof will not require any special framing. If you like tile, don't let its weight deter you.

When considering re-roofing an old house with tile, an owner should consult a building inspector or engineer to see if the structure was designed to handle the weight. While most houses can support a tile roof, an occasional old house that was built to minimum standards may have to be reinforced before tiling. Usually, this reinforcement takes the form of a braced purlin at mid-span, and the structural work adds a few hundred dollars to the

roofing expense. An alternative would be to choose one of the lightweight-concrete or hybrid fiber-cement tiles on the market. At 40% to 60% of the weight of standard concrete or clay tile, these materials generally need no additional bracing. But watch the costs. Bracing can be cheaper than the premium charged for these lightweight materials.

During an earthquake, the rolling and shaking of a house has been known to flip unsecured tiles off the roof. Also, on tall houses especially, the weight of any heavy roofing sets up extra stress in the frame as the house whips back and forth. For those of you building in earthquake country, consider taking the precaution of securing at least every other row of tile with nails. And make sure that the house has enough shear strength to resist wracking. Check with your local building department or consult with a structural engineer if you are in doubt about the ability of your house to bear the weight of a tile roof during an earthquake.

I was curious about the effects of the 1989 earthquake in Santa Cruz, California, on tile roofs, so I called the building department there to find out how tile roofs fared during the shake. I was told that, in general, tile roofs held up well. Any tiles that were shaken loose were unsecured and were typically on roofs steeper than 5-in-12. Tall houses with tile roofs showed no structural damage, reflecting the soundness of the tough engineering requirements that have been in force there for some time.

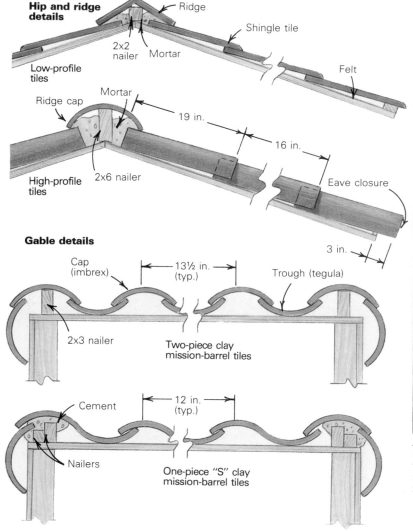

**Hip and ridge details**

Ridge

2x2 nailer

Mortar

Shingle tile

Low-profile tiles

Felt

Ridge cap

Mortar

19 in.

16 in.

High-profile tiles

2x6 nailer

Eave closure

3 in.

**Gable details**

Cap (imbrex)

13½ in. (typ.)

Trough (tegula)

2x3 nailer

Two-piece clay mission-barrel tiles

Cement

12 in. (typ.)

Nailers

One-piece "S" clay mission-barrel tiles

***High-profile clay tiles.*** Tapered two-piece mission-barrel tiles have nail holes at opposite ends, depending on whether they are cap or trough tiles. The cap tiles above are secured with long nails. It's best not to walk on a tile roof—all the cap tiles behind the broken cap in row five will have to be pulled before it can be replaced.

**What will it cost?**—Not surprisingly, the cost of a tile roof will depend on the type of tile, the distance from the manufacturer to your site, the complexity of the roof, the quality of installation you specify, and who installs it. Like other premium roofing products, such as slate and metal, tile costs more initially than an ordinary composition roof but can have a lower life-cycle cost because it lasts much longer. Also, the appreciation in property value and lower maintenance must be figured into the calculation.

Concrete tile generally is the least expensive type of tile roofing. A basic concrete-tile roof installed by a knowledgeable roofer in an area where tile is common, such as the Sun-belt states, will cost about $125 to $150 per square (100 sq. ft.), with about half of that as materials and the rest labor. In an area where tile is still a new idea or with a roofer just getting started in tile, the cost will be higher. A complex roofline will cost more for both labor and materials. Premium flashing and fasteners will also add to the expense.

In the South and Southwest, the home of clay tile, a new roof of unglazed, one-piece mission tile will cost the same or a bit more than concrete. Two-piece mission-barrel tiles run a little more because there are more tiles to install. For most other areas of the country, where clay-tile plants are scarce, shipping adds considerably to the cost.

Glazed clay tiles stand on the top rung of the cost ladder. The tiles alone start at just under $170 per square—about the same as treated shakes—with shipping, fittings, labor and profit on top of that. The most expensive glazed tiles that I've heard about top $1,000 per square. Most people who are contemplating a tile roof cannot justify the additional cost of glazed tile versus unglazed. There are times, however, when least cost is not the object, and intangibles enter into the equation. The rich color and texture of some of the glazes on classic profiles may seduce an otherwise analytical voice that speaks of budgets and mortgages.

**Finding the right tile**—Your first step in selecting a roofing tile is to reconcile your taste with your budget. Once you have decided on a general category of tile, you will want to find a reputable manufacturer. Concrete tiles are now manufactured throughout the Sunbelt states, and plants are beginning to spread across North America and Hawaii. Clay tiles come from plants in California, Florida, Ohio, and New York. Glazed tiles are made in California and Ohio, and are imported from the Orient to West Coast ports. To find a nearby manufacturer, check with your local roofing-supply house. If you cannot find the tile you want, call or write the National Tile Roofing Manufacturers Association (NTRMA, 3127 Los Feliz Boulevard, Los Angeles, Calif. 90039; 800-248-8453). Tell them what you are looking for, and they will direct you to a manufacturer or help you find the tile that is right for your project.

Before you buy, check into the reputation of the manufacturer. If you choose a tile from a member of the NTRMA, the tiles will meet or exceed ICBO (International Conference of Building Officials) standards and have a 50-year warranty. If you choose a tile made by someone else, ask about their warranty and for a copy of their ICBO report. Whatever you do, walk away from a supplier who cannot or will not provide an ICBO report that matches the identifying marks on the tile.

As important as finding the right tiles is finding the right person to install them. The place to begin is with the manufacturer. Most manufacturers keep a list of roofers who have been trained to install roofing tiles. When you talk with the roofer, start out with the standard questions about experience, licenses, insurance, and references. Then together go over a copy of the manufacturer's installation guidelines. If the roofer suggests varying some detail, check that it will not void the manufacturer's warranty. Ask the roofer about upgrading the flashings, fasteners, and felts. It seems false economy to scrimp on the details after committing to the investment in a roof that will outlive the next generation. Here may be the one insoluble disadvantage of a tile roof: it forces us to think beyond our own mortality.□

*J. Azevedo is a free-lance technical writer, working in Santa Clara, California.*

**Glazed clay tiles.** Tiles made of clay can be finished with baked-on glazes in both custom or stock colors. This Japanese-style roof in glossy white is essentially the same as the interlocking S-type mission tile, but its trim pieces, such as the cylindrical hip starters and the belled ends of its hip and ridge tiles, give it an exotic flair.

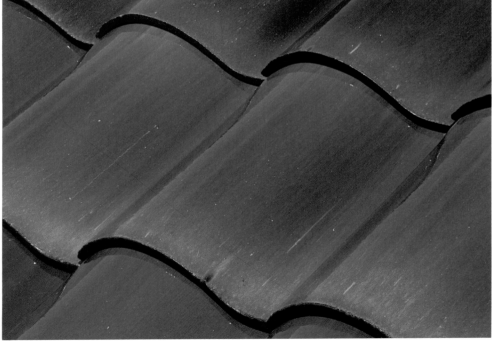

While most tile roofs cap a house in earth-toned silence, the rich colors made possible by using vitreous glazes on ceramic tiles can make the impact of the roof the dominant element in the bearing of a house.

# Roof Shingling

## With only a few rules to follow, putting on a wood roof can be relaxing work with pleasant materials

### by Bob Syvanen

Shingling is one of my favorite tasks in building houses. Even though roofers may be a little faster, I like to do this work myself. Shingling is the kind of job that requires little calculating and a minimum of physical effort. You can think of other things as you work. You don't have to manipulate unwieldy boards or carry heavy loads. Shingle nails are fairly short, so swinging the hammer is easy on the arm; and if you have the time to invest in the old-fashioned methods of making hips, ridges and valleys, there's just enough cutting and fitting to make the job interesting.

Wood shingles are typically three or four times as expensive as asphalt shingles, but they give a roof a texture and color that you can't get with petroleum products. A wood roof is also much cooler in the summer, and will last nearly twice as long as one covered with conventional asphalt shingles. The only major disadvantage to wood shingles is their flammability; but chemical treatments, along with spark arrestors on fireplace chimneys, can minimize this liability.

In the past, shingles were commonly made of cypress, cedar, pine or redwood. My favorite is cypress, although red cedar is what's most available these days. It too is excellent for roofs because the natural oil in the wood encourages water to run off instead of soaking in, and it helps prevent the shingles from splitting despite wide fluctuations in humidity and temperature year after year.

Wood shingles are a delight to work because they're already cut to length and thickness from the best part of the tree, the heartwood. Shingles are sawn flat on both faces, which distinguishes them from shakes, which are split out along the grain. Wood shingles come in lengths of 16 in., 18 in. or 24 in., and taper along their length. The exposed ends, called butts, are uniformly thick for each length category of shingles. This measurement is always given in a cumulative form—16-in. shingles, for instance, always have butts that are 5/2. This means that five shingle butts will add up to 2 in.

**Tools**—Tools for shingling are few and simple. I have put on many shingles using a hammer and a sharp utility knife. Most pros use a lathing or shingling hatchet. A shingling hatchet has an adjustable exposure gauge on the blade; however, a mark on your hammer handle works almost as well. Two good features of the hatchet are the textured face on the crown, and the hatchet blade itself. The mill face or waffle head is less likely to glance off a nail onto a waiting finger, especially when the head of the hammer strikes a blob of zinc that hot-dipped shingle nails often have. The sharp blade and heel of the hatchet are useful in squaring shingles, and in trimming hips and rakes. I also use a block plane to trim hip, valley and ridge shingles for final fit.

I prefer to keep my nails in a canvas apron at my waist, but others like leather nailbags hung off a belt. Production roofers use a stripper, a small, open aluminum box that straps to their chest. It has slots that allow the points of the nails to drop into line when it's loaded with a handful of nails and shaken back and forth.

Although it's possible to work from a ladder, proper staging or scaffolding is a big help when starting a roof. Wall brackets (drawing, facing page, left) are my first choice. You can make them from 2x4s, or rent or buy the sturdier steel ones. They should be attached to the wall studs at a comfortable working height below the starter course on the eave. Some brackets will accommodate nearly 30 in. of scaffolding planks, but two 2x10s battened together make

## *Figuring materials*

You can usually buy shingles in three grades. Always use No. 1, the best grade for a roof. No 1. shingles are 100% clear heartwood and edge grain. No. 2s and 3s have more sapwood, knots and flat grain. They're okay for outbuildings where the life expectancy of the structure is shorter, or for starter coursing, shim stock and sidewall shingling.

In order to figure how many shingles you will need, you must first know what exposure you are going to use. Exposure is the measurement of how much of the shingle shows on each course. The longer the shingle, the greater the possible exposure. Maximum exposures are also determined in part by the pitch of the roof. As shown in the chart below, the flatter the pitch, the less shingle that can be left to the weather.

| Maximum exposure (in.) | | | |
|---|---|---|---|
| Roof pitch | Shingle length (in.) | | |
| | 16 | 18 | 24 |
| 5-in-12 and up | 5 | 5½ | 7½ |
| 4-in-12 | 4½ | 5 | 6¾ |
| 3-in-12 | 3¾ | 4¼ | 5¾ |

Using these maximum exposures, all pitches of 5-in-12 and up will give triple coverage. With lesser pitches, successive courses of shingles will overlap each other four times, so you get quadruple coverage. Wood shingles shouldn't be used on pitches lower than 3-in-12.

When the maximum exposure is used on any length shingle, four bundles of shingles will cover 100 sq. ft. of roof, or one square. To figure your roofing material, multiply the length of your roof by its width and divide by 100. This will give you the number of squares. For starter courses, add one extra bundle of shingles for each 60 lineal feet of eave. A starter course is the first course of shingles, which is doubled to provide a layer of protection at the joints between shingles. In some cases, the starter course is even tripled. For valleys, figure one extra bundle for every 25 ft., and the same for hips. If there is a hip and a valley, figure some of the waste from the valley to be used on the hip. If you are going to use manufactured hip and ridge shingles, a bundle will cover about 17 lineal feet.

When you calculate shingle quantity, figure nails and flashing, and order them at the same time. For 16-in. shingles on a 5-in-12 or steeper pitch, use 3d galvanized shingle nails and figure 2 lb. per square. For 24-in. shingles, get 4d nails, with 5d nails or bigger for re-roofing jobs when you're nailing through other shingles. Hip and ridge caps need nails two sizes larger than the shingle nails used in the field (on the roof slope), because you will be nailing through extra thicknesses.

If you are near a good-sized city, roofing-materials suppliers are fairly common. Their prices are often more reasonable than lumberyard prices, and their inventory is only for roofers, so you're more likely to get the flashings and nails that you need. Ask for the price per square on the shingles that you want, and be prepared to give a figure of how many squares you'll need. Another advantage of buying from a roofing supplier is that many of them deliver on lift-bed trucks and load the roof with the shingles, saving your back and a lot of time walking up and down a ladder. —*Bob Syvanen*

From *Fine Homebuilding* (December 1982) 12:52-57

a nice working platform. Make sure the battens (or cleats) extend back beyond the planks to the wall to prevent them from shifting on the brackets inward under the eaves. Scaffolding planks and steel ladder brackets hung from the rungs of two straight extension ladders placed against the siding will also work nicely. Once up on a roof over a 4-in-12 pitch, I use roofing brackets, but I'll get to these later.

**Preparation**—One decision you make before shingling is what sheathing to use. Where there is no wind-blown snow, or where the weather is humid and wet much of the time, an open slat roof with spaced sheathing is a good choice. Shingles are laid on top of it without roofing felt, so air can circulate freely on the underside of the shingles. The spacing of this sheathing is important. For a 5-in. exposure, the 1x4 sheathing should also be spaced 5 in. o.c. (drawing, below right). The shingle tips should lap over the sheathing at least 1½ in., with two boards butted together at the eaves and ridge for proper nailing of starters and ridge caps.

In snow country, the solid-sheathed roof is best. I have stripped many roofs with solid sheathing and found that they have held up very well. I use CDX plywood and cover it completely with 15-lb. felt. If you are using felt, lay only as much as you need for a day's work. Morning moisture will wrinkle the paper and make shingling difficult. Along the eaves I use 36-in. wide 30-lb. felt. If ice-damming is a particular problem in your area, you can trowel on

**Using a shingler's hatchet, Syvanen squares up a red cedar shingle, right. The 1x3 roof stick he is cutting on is a gauge for laying straight courses with the correct exposure. The shingles on the top course at the left of this photo have been butted to this gauge and nailed, each one with two 3d shingle nails.**

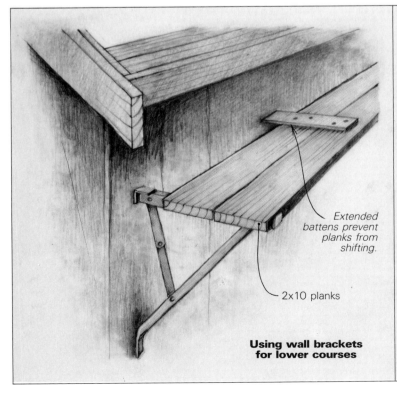

*Extended battens prevent planks from shifting.*

2x10 planks

**Using wall brackets for lower courses**

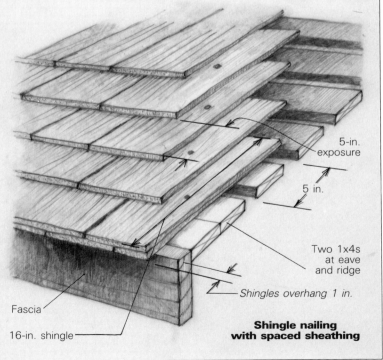

5-in. exposure

5 in.

Two 1x4s at eave and ridge

Shingles overhang 1 in.

Fascia

16-in. shingle

**Shingle nailing with spaced sheathing**

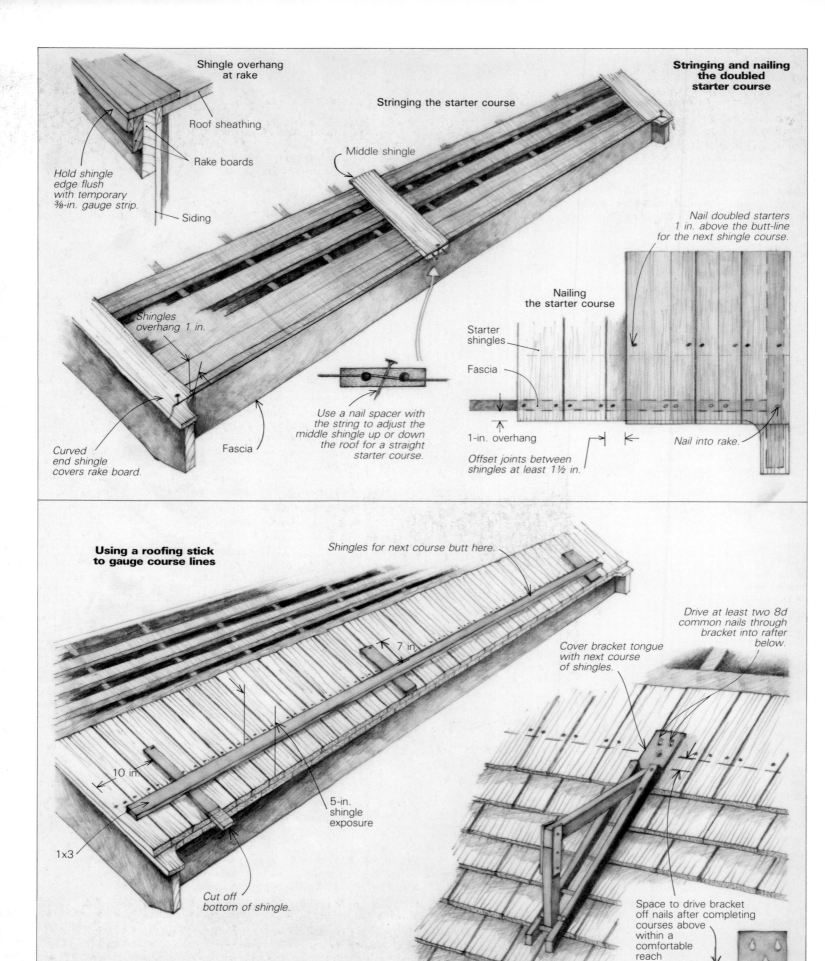

Shingle overhang at rake

Roof sheathing

Rake boards

Hold shingle edge flush with temporary ⅜-in. gauge strip.

Siding

Shingles overhang 1 in.

Curved end shingle covers rake board.

Fascia

Stringing the starter course

Middle shingle

Use a nail spacer with the string to adjust the middle shingle up or down the roof for a straight starter course.

**Stringing and nailing the doubled starter course**

Nail doubled starters 1 in. above the butt-line for the next shingle course.

Nailing the starter course

Starter shingles

Fascia

1-in. overhang

Offset joints between shingles at least 1½ in.

Nail into rake.

**Using a roofing stick to gauge course lines**

Shingles for next course butt here.

7 in.

10 in.

1x3

5-in. shingle exposure

Cut off bottom of shingle.

Drive at least two 8d common nails through bracket into rafter below.

Cover bracket tongue with next course of shingles.

Space to drive bracket off nails after completing courses above within a comfortable reach

**Installing roofing brackets**

Illustrations: Frances Ashforth

When using roof brackets, or jacks, squatting is usually the most comfortable working position. Once you have shingled up the roof to a point where you are working at full arm extension, nail a new line of brackets to the rafters higher up the roof. These brackets are removed when the roof is completed by tapping them at the bottom. The 8d common nails used to secure each bracket remain under the shingles.

a layer of roof mastic over the 30-lb. felt at the eaves, and then lay on another run of felt over this for a self-sealing membrane.

The shingles will need to overhang both the rake and eaves. For the rake overhang, cut some temporary gauge strips ⅜ in. by 1 in. or so, and tack them on the rake board with 4d nails (drawing, facing page, top left). You can then hold your shingles flush with the outside of this gauge board for a ⅜-in. overhang.

**Starter course**—The first course along the eave is doubled. I like to extend the end shingle of this starter course over the rake board or gutter if there is one. Water runoff will wear away the top of the exposed piece of rake board if it is not covered. This extended shingle can be straight or curved. I like it curved. I use a coping saw and cut several at one time.

To line the starter course, nail the curved end shingles in place. These shingles will be flush with the rake gauge strips and will overhang the fascia on the eave 1 in. Tack a shingle in the middle of the run with the same 1-in. overhang. Because the fascia probably won't be truly straight, the middle shingle will have to be adjusted up or down to straighten the starter course. Do this by stretching a string from one end shingle to the other on nails spotted at the line of the 1-in. overhang.

To keep from butting the shingles directly to the string and introducing cumulative error by pushing against it, fix the string away from the line of the starter course by the diameter of a nail. Then use a loose nail to gauge each starter shingle to the string as you nail it. Begin by adjusting the middle shingle to the string in this way. The drawings at the top of the facing page show how to get the string right.

Once the middle shingle is nailed down, fill in the rest of the starters, nailing them all into the fascia. Nail the end shingles into the rake board (drawing, facing page, top right). Angle the nails away a little to make sure that they don't poke through the face of this trim.

On top of the starters, nail the double starter shingles, making sure that the gaps between them are spaced at least 1½ in. from the gaps in the row below. The shingles in this course can overhang the starters by about ⅛ in., or sit flush. They should be nailed about 1 in. above the butt line for the next course; for a 5-in. exposure, nail about 6 in. up on the shingles.

Roof shingles get a lot of water dumped on them, and they will buckle if placed too close. I just eyeball the distance between edges, but any spacing from ⅛ in. to ¼ in. is fine. The joints between shingles should be offset from the joints in the course directly below by at least 1½ in., and offset from the gaps in the course below that one by at least 1 in.

**Roofing sticks and brackets**—To lay the second and all successive courses, you'll need a method of gauging the exposure and keeping the courses straight. One way is to use the gauge on your hatchet or a mark on your hammer handle. I prefer to make and use roofing sticks to align an entire course before I have to reposition them. Use as many of these sticks as it takes to span the roof. To make one, take a long, straight length of 1x3 and nail a 2-in. wide shingle to it at right angles every 8 ft. or so (drawing, facing page, bottom left). Then saw off the butt of the shingle so that the distance from this cutoff edge to the top edge of the 1x3 equals your single exposure. To use the roofing sticks, line up the shingle butts on the gauge stick with the butts of the first course. Tack the gauge to the roof with a nail in the upper corner of each gauge shingle. By butting the shingle ends down against the top edge of the 1x3, you get uniform exposure from one course to the next. On a calm day, you can lay in quite a few shingles before nailing them down—a nice feature of this system.

When you reach one of the shingle tips that is part of the roofing stick, just fit the shingle without nailing it. Tuck it under the butt of an adjacent shingle for safekeeping and continue down the course. Then when you've completed the full course, tap the roof sticks free and nail down the individual shingles you left out.

On even a moderate pitch, you'll need roof brackets to work safely, comfortably and at a good pace. These are adjustable jacks that are nailed to the roof to support a scaffolding plank (drawing, facing page, bottom right). Most lumberyards carry roof brackets, and they can also be rented. The ones I use are oak, but metal ones are more common. The first set can be put on after you've shingled up five or six courses and can't reach across the eave from the ladder or scaffolding.

Make sure your roof brackets are nailed into rafters, and use two 8d galvanized nails per bracket. Most brackets will take a single 2x10 plank. Twelve-foot planks are ideal when you

use a bracket at each end. Fourteen-footers will do. You can shingle right over the tongue of the bracket where it attaches to the roof, because the nails that hold it will be left under the shingles when it's removed. When you install the brackets, make sure there's enough space between the top of the bracket and the shingle butts to allow you to remove the bracket easily, by driving its bottom toward the ridge, and lifting the tongue off the nails.

**Do's and dont's**—Face-grain shingles have a right and wrong side up. Make sure to place the pith side of the shingle (the face that was closer to the center of the tree) down to prevent cupping. Some shingles are cupped or curled at their butts. I put the cup down for better runoff and appearance. I reject hard shingles because they split when nailed and curl up when the sun hits them.

Red cedar shingles can be very brittle, so hold back on that final hammer stroke or the shingle may crack. The nail head should sit on top of the shingle, and not be driven flush. Each shingle, no matter how wide, should get just two nails. If you have a hip or valley to shingle, save the bedsheets (shingles 9 in. or wider) when you're roofing in the field.

If you use a wide shingle in a regular course, score it down the middle with a hatchet or utility knife to control the inevitable splitting. Treat it as two separate shingles, offsetting the scored line from other joints, and nailing as you would two shingles. If you suspect that a shingle is checking, bend it into a slight arch with your hands. If it has a clean crack, use it as two shingles. If it shows several cracks, throw it out.

When you are shingling in the field, don't get so tightly focused on the work that you forget to check every once in a while that the coursing is even and straight. Do this by measuring up from the eave, and adjust if necessary. You may have to snap a chalkline occasionally to straighten out a course.

Once you've completed the roof, spend a few minutes looking for splits that you missed that

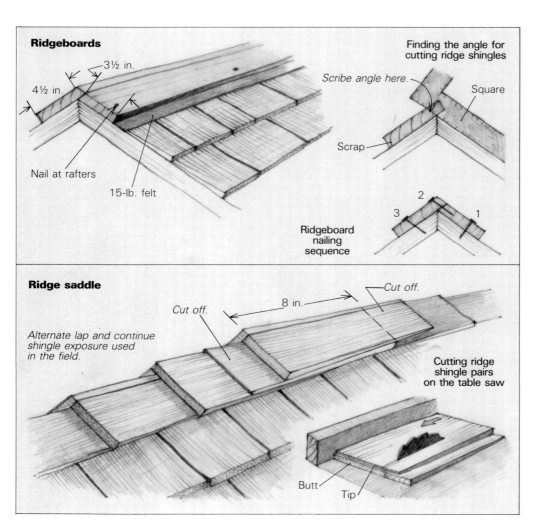

**Ridgeboards**

4½ in.

3½ in.

Nail at rafters

15-lb. felt

Finding the angle for cutting ridge shingles

Scribe angle here.

Square

Scrap

Ridgeboard nailing sequence

2

3    1

**Ridge saddle**

Alternate lap and continue shingle exposure used in the field.

Cut off.

Cut off.

8 in.

Cut off.

Cutting ridge shingle pairs on the table saw

Butt

Tip

are not sufficiently offset from joints between other shingles. Cut some 2-in. by 8-in. strips of flashing out of zinc, aluminum or copper, and slip them under the splits.

**Ridges**—Make sure that the courses are running parallel with the ridge before getting too close to the top. Since a course of 3 in. or less doesn't look good at the ridge, measure up from the course you are working on to the point where the ridgeboard or ridge shingles will come, and adjust the exposure slightly so that the courses work out. If you are using ridgeboards, staple 15-lb. felt over the ridge for the full length of the roof. The felt should be an inch or so wider than the ridgeboard, and will make painting the final coats on the ridgeboard easier. Make your ridgeboards from 1x material, and bevel the top edge of each board at an angle that will let them butt together, and form a perfect peak. Because the boards butt at the ridge rather than miter, one side will be narrower than the other (drawing, top left).

To find the bevel angle for your roof on the table saw, take a pencil, a scrap of wood and a square up to the ridge. Lay the scrap flat on the roof on one side of the ridge so that part of the scrap projects over the peak, as shown in the drawing. Then lay the square flat on the other side of the roof so that it projects past the scrap. Scribe a line onto the wood by holding the pencil against the square. By tilting your table-saw blade to this angle, you will be able to rip both the narrow and wide ridgeboards.

To install the ridgeboards, snap a chalkline

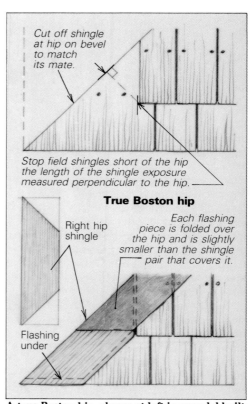

Cut off shingle at hip on bevel to match its mate.

Stop field shingles short of the hip the length of the shingle exposure measured perpendicular to the hip.

**True Boston hip**

Right hip shingle

Each flashing piece is folded over the hip and is slightly smaller than the shingle pair that covers it.

Flashing under

A true Boston hip, shown at left in a model built by Syvanen, uses the hip shingles to complete each course. The usual way is to superimpose a line of ridge shingles on the hip. The butts of Boston hip shingles can run parallel to the eaves, as shown on the right side of the model, or perpendicular to the hip (left side).

the length of the roof where the lower edge of the narrow ridgeboard will be nailed. Using 8d common galvanized nails, attach the narrow ridgeboard to the roof along the chalkline. Be sure to nail into the rafters. Now nail the wide ridgeboard to the narrow one. This will give a nice straight line at the peak. Lastly, nail the lower edge of the wide ridgeboard into the rafters. Push down on the ridge pieces to force the narrow board against the roof. You might have to stand on the ridgeboards to do this. After the final coat of paint is on, trim off the 15-lb. felt with a sharp utility knife.

You can also use shingles for the ridges—either the factory-made kind that come already stapled together as units, or your own, cut on a table saw. To make your own, set the blade-to-fence distance for the width of your exposure, and set the angle of the blade as explained for cutting ridgeboards. Saw the shingles in alternating, stacked pairs. The bottom shingle in the first pair can have the butt facing the front of the saw table, and the top shingle with the tip doing the same. The next pair of shingles should be reversed so that the tip is on the bottom, and the butt is on top. This will give you shingle pairs whose laps alternate from one side of the ridge to the other.

To install the ridge shingles, staple a strip of 30-lb. felt down the full length of the ridge. In this case, the paper should be narrower than the ridge unit. Nail the ridge pieces along a snapped chalkline using a double course as a starter. Alternate the pairs and use the same shingle exposure you have on the rest of the roof. Again, the nails should be about 1 in. under the butt of the next ridge shingle. When you reach the center, make a saddle by cutting the tips off two pairs of ridge shingles so that 8 in. remain on each (drawing, facing page, center left). Then nail them down on top of each other with their butts facing the ends of the ridge.

**Hips**—You can buy factory-assembled ridge units for hips, or make your own. Most folks just staple down a run of 30-lb. felt over the hip and nail the units along a snapped chalkline or temporary guide boards. Since water drains away from a hip, this method keeps the rain out. But when I have the time, I like to cut and fit a true Boston hip the way the old-timers did (photo and drawing, facing page, bottom). It weaves the hip shingles and flashing right into the courses in the field for a weathertight fit that doesn't look added on, like standard hip units. It looks tough to do, but it's not really.

The key to working a true Boston hip is to stop the shingles in the field the same distance from the hip on each course. This makes the hip shingles uniform, allowing you to cut them on the ground. To find the point on a course line where the butt of the last shingle should end, measure the length of the shingle exposure you are using in the field (such as 5 in.), on a line that runs perpendicular to the hip. You will have to move this 5-in. line, marked on a tape measure or square, up or down the roof (keeping perpendicular to the hip) until it fits between the hip and the course line. Then

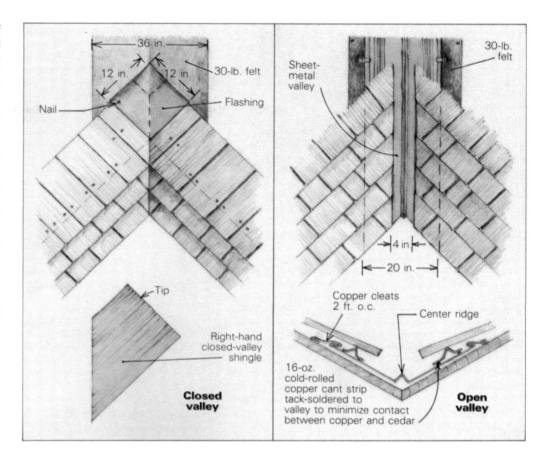

mark the intersection on the course line and fill in the shingles in the field to that point.

You will need to make right and left hip shingles to form pairs. The tip of each hip shingle needs to be trimmed to fit into the space left for it by the last shingle in its course and the one above it. The butts can be left perpendicular to the hip, or cut parallel to the eave. The long side of the shingle also needs to be cut. This should be a bevel that alternates lapping its mate on the other side of the hip. You can cut these bevels on a table saw or use a shingling hatchet on the roof.

Flashing pieces, which are used under every hip shingle, should be cut out with snips on the ground. If you put a slight crease down the center of each piece, it will straddle the hip easily. The flashing pieces should be slightly narrower than each hip unit so they don't show, yet long enough so that each piece laps the previous one by a good 3 in.

**Valleys**—It's critical that valleys get done properly, since the roof directs water right to them. Valleys can be open or closed (drawings, above). The open valley can be shingled faster and cleaned more easily. In closed valleys, leaves and pine needles can be a problem, and some people don't recommend them. But I prefer their neat look. Shingles butt up tight in the valley with each course flashed much like Boston hip flashing.

For a closed valley, start with a 36-in. piece of 30-lb. felt laid in the full length of the valley, then add a piece of 12-in. by 12-in. flashing cut diagonally in half. Add the starter course of shingles. Next comes a 12-in. square piece of flashing, and the second course of shingles.

Each valley shingle must get a miter cut

along its inner edge where it meets the other valley shingle. It is often easier to cut these on the ground, and do your final fitting on the roof with a hatchet or block plane. Use the bedsheets that you saved. If you are cutting ahead of yourself on the ground, don't forget to cut both right and left-handed shingles. The flashing pieces can also be cut ahead. Remember when you nail both shingles and flashing to use only one corner nail as far away from the center of the valley as you can on each side.

For an open valley, also lay in a 36-in. wide sheet of 30-lb. felt; it's a good bed for the metal that follows. On pitches under 12-in-12, use 20-in. wide sheet metal, which should be nailed at the extreme edges with fasteners that are compatible with the flashing (see *FHB* #9, p. 50). Tin, lead, zinc and galvanized steel are all right for valley metal, but I like to use copper. There have been a few cases of copper flashings and cedar shingles reacting chemically to produce a premature corrosion of the copper. I have never seen this, nor has anyone I know. Just to be safe, you can use a cant strip to minimize the contact between the shingles and the copper (drawing, above right). Whatever flashing you use, it will be improved by a crimp in the center.

I often begin in the valley by nailing the first few shingles on five or six courses high before I carry the courses across the roof. This stacking allows me to set the valley shingles to a chalkline without being restricted by having to fit them to the shingles in the field. Just as with closed valleys, it is important to nail as far from the valley center as possible. □

*Bob Syvanen is consulting editor to* Fine Homebuilding, *and a builder on Cape Cod.*

# Roofing with Asphalt Shingles

## There's more to laying three-tab shingles than just nailing them on as fast as possible

by Todd A. Smith

I guess I was destined to work for my father's 63-year-old roofing company. I spent a lot of time on the roof learning the trade from several old craftsmen. I learned slate roofing, tile roofing, copper roofing and, of course, asphalt-shingle roofing. The most important lesson I learned was that there's more to installing shingles than just nailing them on fast. As a roofer, I am also charged with preventing leaks, making a house more attractive and re-

membering that everything a person has worked hard for is under my roof. In this article I'll describe the basics of installing a tight, durable three-tab asphalt-shingle roof.

**Tools**—You don't need many tools to install a shingle roof. A hammer, tin snips, utility knife, tape measure and chalkline will do it. I use a drywall hammer to nail shingles. Its light weight doesn't tire my arm, and the larger

head makes nails an easy target. To carry nails and tools, I wear a leather carpenter's apron, with suspenders to support the weight. Cloth aprons seem to wear out too easily. Besides, I like the extra pockets in my leather apron for different nails.

I prefer the snips when I have to cut shingles that butt into flashing or siding because I can make more accurate and intricate cuts with them. Otherwise, I use a hook blade in

**With a column of shingles run up the center of the roof to the ridge, one roofer works to the left of center, the other to the right. This roof is shallow enough to walk on, but the roof brackets and scaffold provide a little extra room to stand and a place to rest bundles of shingles. Looped around the brackets, lengths of chain with hooks on the end make a handy place to hang a nailer or a bucket full of nails.**

my utility knife because it's less likely than a straight blade to cut whatever is underneath the shingle. When cutting a shingle on the roof, I use the back of another shingle as a straightedge, which saves me from having to carry a square.

Our company's air-powered nailers speed up production, but they have their drawbacks. Nailers are heavy, and dragging 100 ft. of hose is cumbersome. Our nailers hold only enough nails to install one bundle of shingles, which means we have to keep coils of nails on the roof, and they get in the way. Also, in areas that require lots of cutting and fitting of shingles, power nailers are clumsy and impractical.

To make the nailers easier to handle on the roof, I made some portable utility hooks that attach to our scaffold brackets. This gives me a place to hang my nailer when I'm not using it (photo facing page), as well as a place to hang buckets of nails and coils of air hose. The hooks are made with loops of chain, spring latches and some utility hooks, all of which I got at a hardware store.

**Ladders and roof brackets—**For us, safety begins as soon as the trucks show up at the job. Roofing can be hard on your back, and you have to be careful just pulling the ladders off the truck and setting them up, not to mention hauling heavy rolls of felt and bundles of shingles up on the roof.

I've become accustomed to setting up extension ladders alone because I have to examine many roofs by myself. I slide the ladder off the back of the truck until its feet touch the ground, then tip it up on one edge. Squatting under the ladder with a little over half its length in front of me, I stand up, placing the ladder on my shoulder, keeping my back straight and lifting with my legs. With the majority of the weight in front of me, I don't have any downward pressure behind me to strain my back.

To stand the ladder up, I set the foot of the ladder against a solid object—usually a foundation, step or tree trunk—and push on the top, walking my hands down the rungs. Once the ladder is straight up, I raise it to the appropriate height, watching out for power lines, phone lines and tree branches.

There are two accessories I use with my ladder when the needs arise. One is a ladder standoff, which is a large U-shaped affair that bolts to the ladder and prevents it from leaning directly against the eaves of the house. The other is a ladder scaffold, which is a platform supported by two brackets that hang over the rungs on a pair of extension ladders.

I use the ladder scaffold to work along the eaves of a house. My steel scaffold platform is 20 ft. long, and to be sure it overhangs the brackets at least 12 in., I stand up two ladders about 17 ft. apart. Once the ladders are up, I set the brackets at the proper working height, which for me is about 3 ft. below the eaves. The brackets are adjustable and can be oriented in a horizontal position no matter what the ladder's angle. Because of its size and weight,

getting the 20-ft. scaffold up the ladder and onto the brackets is a two-man job, and once in place, I tie the scaffold to the ladders.

About 90% of the time I'm working on a roof that's too steep to walk on (anything greater than a 6-in-12 pitch), so I set up roof brackets (photo facing page). These are triangular steel brackets, some of which can be adjusted level regardless of the roof pitch. Others are fixed and create a working surface that allows me to walk around on the roof easily (for more on scaffolding see *FHB* #36 pp. 34-38).

**The basic materials—**Asphalt shingles and fiberglass shingles are the most common roofing materials used today. Both are less expensive than slate, tile or metal roofing and are more fire resistant and maintenance-free than wood. The main difference between asphalt and fiberglass shingles is in the mat, or base sheet, that manufacturers begin with when they make shingles. Asphalt shingles have an organic base, which, like felt underlayment, is

composed of cellulose fibers saturated with asphalt. Fiberglass shingles have a base of glass fibers and don't need to be saturated with asphalt.

Beyond that, asphalt and fiberglass shingles are made pretty much the same way. The mat is coated with asphalt on both sides, and then ceramic-coated mineral granules are applied. The granules help shield the asphalt from the sun, provide some fire resistance and add color. The seal strip, applied along the width of the shingle and just above the cutouts, is activated by the sun and seals each shingle to the one below it.

On a standard three-tab shingle, the top edge of each shingle is marked with a notch every 6 in. The notches are used to register the shingles in the next course, since alternate courses are offset from each other by 6 in. Some manufacturers even notch the shingles every 3 in. to allow for a pattern that repeats every third row as opposed to every other row.

Fiberglass shingles are a little more difficult to install than asphalt. In hot weather, they become softer much faster, which makes them harder to handle and cut, and easier to damage.

Fiberglass shingles are lighter in weight, though. A square (100 sq. ft.) of standard three-tab fiberglass shingles weighs 225 lb. A square of asphalt shingles weighs 240 lb.

The two most common types of roofing felt are 15-lb. felt and 30-lb. felt. The designation is based on the weight of a 100-sq. ft. area of felt. Although some people don't bother to install felt under the shingles, we always do. For one thing, felt protects the roof decking from the weather. Even though roof decks are made from exterior-grade plywood, they won't withstand prolonged exposure to the elements. Second, roofing felt helps prevent ice and snow from backing up under the shingles and leaking water into the house.

**Installing felt and flashing—**Felt is applied in courses, parallel to the eaves. Generally courses overlap each other by 4 in. The rolls are marked with white lines along the edges to help you maintain a consistent overlap. There's also a pair of lines in the center of the felt, in case you want to overlap the courses 18 in. (half the sheet). You might want to do this on a shallow-pitched roof, say 4-in-12, in an area prone to ice damming.

We try to run the length of the roof with a piece of felt, trimming it flush with the gable ends. But if we have to splice in the middle of a course, we overlap the ends 4 in. We run felt 6 in. over all hips and ridges (from both sides). Valleys are lined with a full width (36-in.) piece of felt first, and then the courses are run into it, overlapping the sides of the valley felt by 6 in. Where a roof butts into a sidewall, we run the felt 4 in. up the wall.

When we come to a vent pipe, we cut a 3-ft. piece of felt, make a hole in it the size of the pipe, slip it over and seal around the pipe with roofing cement. Then, we overlap the felt on both sides of this piece. To avoid having to chalk horizontal lines later as a guide for the

## Estimating roofing

Roofing shingles are sold by the square: a square of shingles is enough to cover 100 sq. ft. There are 27 shingles in a bundle of standard three-tab shingles, and three bundles to a square. I calculate the square footage of the roof and divide that figure by 100 to determine how many squares of shingles I need. For every 5 ft. of hip or ridge, I need four shingles. I generally figure 30 ft. to 40 ft. of hips and ridges for every square of shingles. When a roof has many hips, valleys and irregular shapes, I figure 10% to 20% extra for waste.

From the blueprint or specs, I also determine if I will need drip edge, flashing or any other additional material. The drip edge and flashing are linear dimensions. I figure 2 lb. of nails for each square of shingles. I use 1¼-in. galvanized nails for new construction. I don't use staples to install shingles because I feel the head of a nail holds the shingle on better. Manufacturers of roofing felt assume a 2-in. overlap, so a roll of 15-lb. felt that will cover 400 sq. ft. of roof has 432 sq. ft. in the roll.

The cost of installing the roof is the sum of the materials, labor, overhead and profit. Most roofers determine the labor cost based on a set price for each square of roofing. The problem with this is that not every square of roof shingles takes the same amount of time to install. A square of roofing material on a shallow-roofed ranch house will take less time to install than a square of roofing three stories high on a steeply pitched Victorian. Instead, I break down the roof into sections and determine the time each one will take a specific roofer or group of roofers to complete. I am always conscious of details that require extra time, like valleys and flashings. —*T. S.*

shingles, we install the felt as straight as possible so that we can measure off of it to keep the shingles straight.

We nail along the seams and edges of the felt. In the center of each course, we nail every 2 ft. or so. Although you can nail by hand, I usually use a power nailer filled with 1¼-in. roofing nails, which are the shortest pneumatic nails I can get. When I need to install a small area of felt quickly and don't want to bother with a compressor and hoses, I use an Arrow Hammer-Tacker (Arrow Fastener Company, Inc., 271 Mayhill St., Saddle Brook, N. J. 07662). It's a staple gun that is used like a hammer, but tends to gum up with felt after a lot of use. When that happens, we soak the heads in kerosene to break down the tar. Then we scrape them clean and spray them with lubricant.

Once the felt is on, we nail strips of lath along the edges of the roof and along the seams in the felt to prevent the felt from blowing off. Until the roof is done, the felt is the only material keeping the house dry.

The edge of the roof sheathing should always be protected from the elements. I use a metal drip edge, installed with roofing nails, along the gables and eaves. At inside corners, I cut the vertical face of one piece 1 in. long and bend it around the corner, and then cut the other piece so that it butts into the corner. I cut the tops long on both pieces and overlap them in the valley. At outside corners on a hip roof, I cut a "V" out of the top section and simply bend the drip edge around the corner. Gable ends are cut flush with the rake board at the bottom. At the peak, I cut a "V" out of the vertical flange and bend the top section over the ridge. Any splices in the drip edge are overlapped 2 in.

Houses without overhanging eaves are particularly susceptible to damage from ice damming. On such houses we also install a 36-in. wide strip called an ice shield, or eave flashing. Although you can use roll roofing, we use any of several membrane products specifically designed for this like Ice & Water Shield (W. R. Grace & Co., 62 Whittemore Ave., Cambridge, Mass. 02140) or Weather Watch (GAF Corp., 1361 Alps Rd., Wayne, N. J. 07470-3689). We roll them out along the eaves, tacky side down, and nail them only across the top.

**Loading the roof**—Shingles should be stored in the shade, or covered with a light-colored tarp. Otherwise, the sun will heat them up and seal them to each other. For this reason we seldom unload shingles directly onto the roof, even though some suppliers have lift trucks that make it possible. We either carry the bundles up by hand, or we use a gas-powered hoist (ours was made by Louisville Ladder Division, Emerson Electric Co., 1163 Algonquin Pkwy., Louisville, Ky. 40208).

Some roofers carry the shingles up to the ridge and lay the bundles over it. This isn't a good idea, especially with fiberglass shingles. The shingles can heat up, take a set

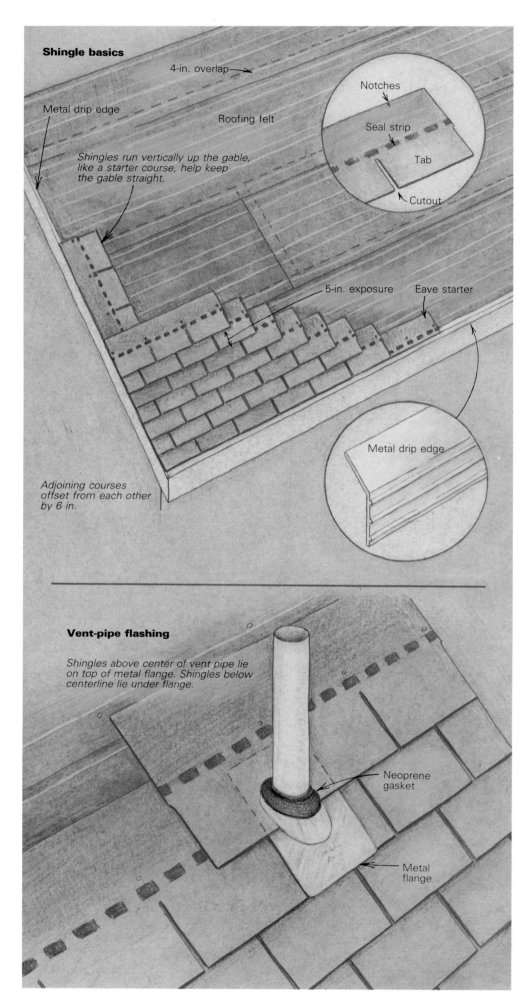

**Shingle basics**

4-in. overlap

Metal drip edge

Roofing felt

Notches

Seal strip

Tab

Cutout

Shingles run vertically up the gable, like a starter course, help keep the gable straight.

5-in. exposure

Eave starter

Metal drip edge

Adjoining courses offset from each other by 6 in.

**Vent-pipe flashing**

Shingles above center of vent pipe lie on top of metal flange. Shingles below centerline lie under flange.

Neoprene gasket

Metal flange

Drawings: Chris Clapp

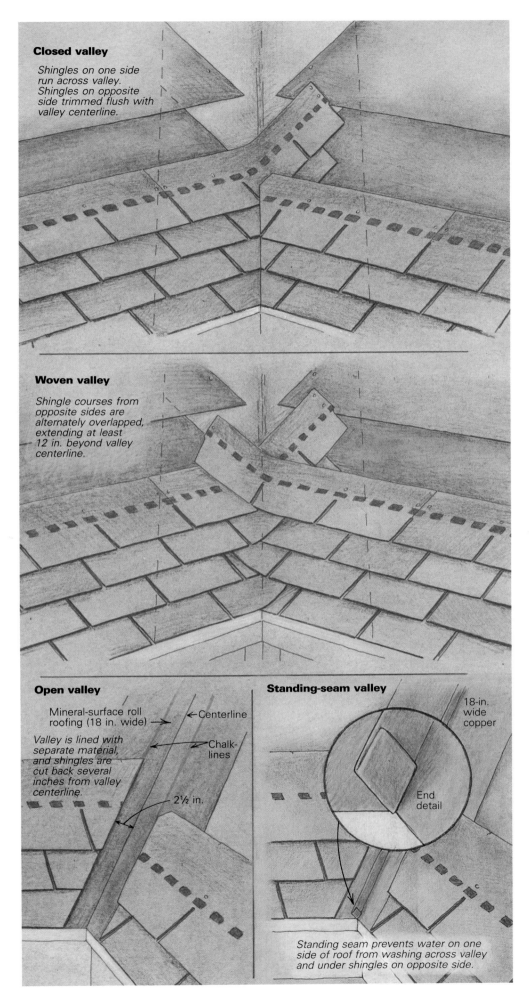

**Closed valley**

*Shingles on one side run across valley. Shingles on opposite side trimmed flush with valley centerline.*

**Woven valley**

*Shingle courses from opposite sides are alternately overlapped, extending at least 12 in. beyond valley centerline.*

**Open valley**

Mineral-surface roll roofing (18 in. wide) → ← Centerline

*Valley is lined with separate material, and shingles are cut back several inches from valley centerline.*

← Chalk-lines

2½ in.

**Standing-seam valley**

18-in. wide copper

End detail

*Standing seam prevents water on one side of roof from washing across valley and under shingles on opposite side.*

from being bent and then not lie properly. We spread the bundles out around the roof, on approximately 6-ft. centers. Any farther apart and you'll work yourself to death. Even on shallow roofs, the bundles can slip on the felt and slide off, so we drive a pair of 10d nails into the sheathing and set the bundles above them. The nails can make a dimple in the edge, especially in hot weather, so we always lay the bundles with top edge of the shingles against the nails.

**Starting the shingles**—Before the first course of full shingles goes on, starter shingles are installed along the eaves. Their purpose is to protect the roof under the cutouts and joints in the first full course and to provide a seal strip for the tabs. Although some people use whole shingles installed upside down as starters, I use full length shingles with the tabs cut off (top drawing, facing page). This puts the seal strip in the right place to hold the tabs.

Along the gable end, I like to let the shingles overhang the drip edge by 1 in. This keeps water from blowing under them. To keep this line of shingles ends straight, I run a course of shingles vertically up the gable, like a starter course (top drawing, facing page). I lay them end to end from eaves to ridge, with the cutouts toward the roof and the top edge overhanging the drip edge about one inch.

Once the starter course is in place, the field shingles can be started. I begin in the center of the roof and work outward, if there is a large open area, or if I'm on a hip roof. I begin on the gable end when there is no large open area to nail shingles, or when there are dormers in the way and I won't be able to get a proper chalkline on the roof.

I install the first five or six courses without chalking any lines, following the eaves with the first course, and then following the notches and cutouts of the previous course. If I need to install roof brackets, I usually do so after the sixth course. This puts them at a comfortable height from which to get on and off the ladder. Also, I can nail two or three more rows of shingles beyond the roof brackets while still standing on the ladder, which is easier than nailing these rows from the scaffold.

Once you're up on the roof, you can work horizontally across the roof, installing several courses as you go. Or you can work vertically up the roof, installing the first two or three shingles in each course all the way to the ridge. I prefer a combination of the two. On our crews one roofer sets the pattern by nailing vertically up the roof (in the center when possible). The other roofers nail horizontally, one to the left and one to the right (photo, p. 100). This increases productivity by allowing three roofers to install shingles on the same roof without getting in each other's way.

We chalk lines on the roof to help us keep the shingle courses aligned and straight. I start with a pair of vertical lines, one in the center of the roof and another 6 in. away (on either side). All subsequent courses are begun from these lines.

If I'm starting from the gable end rather than in the center of the roof, I don't chalk any vertical lines as long as the gable is straight. Before I go up on the roof, I make a pile of full shingles and a pile of shingles that have 6 in. cut off of them. Then when I'm on the roof, all I have to do is alternately pull shingles from each pile and work my way up the gable. Successive courses are automatically offset by 6 in.

When you reach the opposite gable with a course, the shingles are simply trimmed flush with the starter course running up the gable. Rather than trim the last shingle in each course individually, some roofers let these shingles run long and trim them later. If the shingles are not cut as they're installed, I insist that they be cut at least every few hours. If left any longer, they will droop against the side of the house, making it difficult to get a straight cut.

Unless I installed the felt and know that it's straight, I chalk horizontal lines every six to eight courses. (Roofing felt is often installed by the frame carpenters before we get to the job.) I measure 5 in. for each course, which is the amount of shingle exposure (this varies with the manufacturer, so be sure to check this by measuring the depth of the cutout on your shingles). I periodically measure from my chalklines to the ridge to make sure I'm running parallel to it. If not, I adjust the chalklines in small increments.

These horizontal lines help align the top edges of the shingles. Yet because individual shingles may vary in height as much as ¼ in., aligning the top edge doesn't guarantee a straight course. But chalking lines for the top edge means that we can chalk all the lines at once, which saves us time.

If the courses are running off and we need to straighten them out, we'll chalk a line along the tops of the cutouts of the last course installed. This aligns the bottom edges of the shingle and results in a perfectly straight course. If the shingles are running way off, we'll straighten them out gradually over several courses. We always use blue chalk, which washes away in the rain. Red chalk can permanently stain the shingles.

I use four nails to a shingle—one at each end and one over each cutout. On steep roofs I may install a fifth nail at the top center of the shingle. Contrary to conventional wisdom, I always nail right above the seal strip, not on it or below it. Nailing on the seal strip prevents proper adhesion. Below the seal strip, the nails are too close to the edge and can rust or corrode, causing the shingles to slip and the roof to leak. I have never had a shingle blow off or slip because I nailed above the seal strip.

Field shingles that end up over a hip are just run long, nailed and then cut down the center of the hip with the hook blade of a utility knife. At the ridge, we trim one side of the roof flush and fold the shingles from the

Working his way up an open valley lined with mineral-surface roofing, Smith uses tin snips to trim shingles flush with the chalkline. Blue chalk is preferred because it washes away in the rain; red chalk will permanently stain the shingles.

other side over the top. This is just a temporary measure to seal the ridge until the caps are installed.

**Valleys**—There are three different types of valleys used on shingle roofs: the closed valley, woven valley and open valley (drawings previous page). A closed valley is created by running shingles from one side of the roof across the valley (at least 12 in. beyond the centerline), and then overlapping them with shingles from the other side, trimmed flush with the centerline of the valley. It is the cheapest, easiest and least durable option.

The woven valley is created by alternating and overlapping each row of shingles from the left and right sides of the valley. The shingles should overlap the valley by at least 12 in. and nails should be kept at least 8 in. away from the valley centerline.

An open valley is created by lining the valley with a separate material (roll roofing or copper, for instance), and then cutting the shingles back so the lining materal remains exposed. We use open valleys on 90% of our roof installations because they are the most durable. An open valley allows water to flow easily off the shingles. Depending on the budget, the type of shingle and the style of the house, we'll line open valleys with mineral-surface roll roofing, copper or lead-coated copper. One material we never use for valleys is aluminum because it expands and contracts more than other materials, which causes it to wear out much faster.

For mineral-surface valleys, I use strips of material 18 in. wide and 8 ft. to 10 ft. long. A longer piece is difficult to work with and may rip during installation. It comes in 36-in. wide rolls, which I cut down the middle. I cut the valley material on the ground, using a utility knife, and loosely roll it up to make it easier to carry on the roof. Some manufacturers produce roll roofing in colors to match their

shingles. But both black and white roll roofing are pretty common and look good with most shingles.

To install the valley, I start from above and unroll the material. I center it, nail one of the upper corners, then work my way down, nailing the same side every 12 in. to 18 in. It is imperative for the valley material to be tight against the sheathing. If it is not, it can break under the weight of ice and snow. To avoid misalignment, only one person should nail the valley, always working from top to bottom. First one side, and then the other. When I need a second piece to complete the valley, I overlap the first piece 4 in., but don't usually seal between them.

Metal valleys are installed pretty much the same way. Again, I use 8-ft. to 10-ft. lengths. I like to crease the center of the valley metal on a sheet-metal break. The job looks neater and cleaner this way.

I use a standing-seam valley when a steep roof drains onto a flatter shingle roof. I bend the valley metal on a sheet-metal break, creating a ridge in the middle of the valley (bottom right drawing, previous page) that prevents water from the steep side of the roof from flowing across the valley and running up under the shingles of the shallower roof. To finish off a standing-seam valley at the bottom, I cut the sides flush with the eaves, but let the standing seam run about 1 in. long. Then I simply fold the seam back on itself and crimp it tightly. At the top, I start about 6 in. from the ridge and bend the standing seam over with a rubber mallet so that it lies flat across the ridge.

It is important to seal valleys at the top to keep water from getting under them. With a mineral-surface valley, I use a 4-in. wide piece of roofing fabric and apply one coat of roofing cement under it and another coat on top of it. Roofing fabric is a cheesecloth-like material or fiberglass mesh, saturated with

Using a hook blade in a utility knife, Smith cuts ridge caps on the ground (photo left), tapering the tops so that, once installed, the lap portion will be neatly hidden beneath the exposed portion of the succeeding shingle. The caps are centered on the ridge and held in place with one nail on each side (photo above). The last cap on the ridge will have the lap portion trimmed off, and because the nails will be exposed, they'll be sealed with caulk.

asphalt. It comes in 4-in., 6-in. and 12-in. wide rolls. The 4-in. rolls cost about $7 for 100 ft. of fabric and are available at most building-supply stores.

When I don't have anything to solder to, I use the same technique to seal the tops of copper valleys. However, when two copper valleys meet at a ridge or when a copper valley meets copper flashing, I prefer to solder them. To solder two valleys at a ridge, I end one valley at the ridge and bend the other valley over the ridge about 1½ in. (again using a rubber mallet).

After installing the valley material, I chalk lines along both sides of the valley, 2 in. to 2½ in. from the center. These are the marks I follow to install the shingles. The exposed sections of my valleys are between 4 in. and 5 in. wide total. Anything smaller is impractical and anything larger doesn't look good.

I prefer to cut the shingles even with the chalklines while they are being installed, using a pair of tin snips (photo facing page). I nail the full shingle in place, use another shingle as a straightedge, scribe a line with the point of a nail, then bend the shingle up and cut it with my tin snips. I can install and trim the valley shingles on both sides as I work up the valley to the ridge.

Some roofers let all the shingles run long. As each side is completed, they chalk lines down the valley and use a utility knife to cut the shingles (being very careful not to cut the valley material). With either method, it's important to keep the nails 4 in. to 6 in. away from the edge of the shingles in the valley.

**Flashing**—I use step flashing along the sides of a wall or chimney (wood or masonry). Step flashing consists of small squarish pieces of metal, bent in an L-shape. The individual pieces are installed with each course of shingles so that the shingles in the succeeding course hide the exposed metal (for

more on installing step flashing see *FHB #35* p. 50). You can buy precut pieces of aluminum step flashing, but I prefer copper, so I have to make my own from 18-in. wide rolls. I make my steps at least 9 in. wide by 8 in. long, which is equivalent to 5 in. of exposed shingle and 3 in. of headlap. I nail the pieces of step flashing on sidewalls before the siding is installed. Then the siding acts as the counterflashing. In the rear of a chimney, I install a copper cricket, which is a saddle that diverts water around the chimney (see the article on pp. 111-113).

To flash around soil and vent pipes on most houses, we use a manufactured metal flange with a neoprene gasket (bottom drawing, p. 102). It's important to remember that all soil and vent-pipe flashings lie on top of the shingles from the center of the pipe forward and lie under the shingles from the center of the pipe back. We shingle up to the center of the vent pipe, either by notching the shingles around it, or by actually cutting a hole in one of the shingles and slipping it over the pipe. Next we slip the metal flange over the pipe and nail the top corners. Then we continue applying shingles, notching them around the pipe and being careful to keep nails away from the pipe (for more information on flashing techniques, see the article on pp. 106-110).

**Capping hips and ridges**—Hips and ridges are capped with shingles that are only one tab wide. The caps are made by cutting three-tab shingles into three pieces (photos above). Once again, this is done on the ground. The shingles are cut with a utility knife, starting at the top of the cutouts and angling slightly inward, toward the top of the shingle. This assures that the lap portion of the shingle will be neatly hidden beneath the exposed portion of the succeeding shingle (photo above right).

Of all the areas where I have installed shingles, the hips are the most difficult to keep straight without a chalkline. I snap a line parallel to the hip, about 6 in. away (it doesn't matter on which side). This acts as a guide for the outer edge of the hip caps. I start at the bottom, cut the first cap even with the eaves shingles, then I work my way to the ridge.

I nail hip and ridge caps on or just above the seal tab. Nailing higher will cause the bottom of the shingle to pop up. Hip and ridge caps are nailed with about a 5-in. exposure. The bottom of each cap is aligned with the top of the cutout on the previous cap.

Most roofers install ridge caps from one end of the roof to the other, orienting them so that the prevailing winds blow over the caps, not under them. I prefer to work from both ends toward the center. I can't really say why, except that it's the way I was taught. On steep roofs, where the centerline of the ridge is more distinct, I don't usually chalk a line for the ridge caps. On shallow roofs, I do. Once both sides of the ridge reach the center, the last ridge caps have to be trimmed so that they butt together, otherwise the ridge will have a lump in it at this point. One last cap piece will cover the butt joint. It should be only about 5 in. long and installed with two exposed nails, both of which are caulked with clear silicone caulk.

Where the ridge of a dormer meets a roof, I work from the front of the dormer back toward the roof to install the caps. The last cap spans across the seam between the valleys (which is sealed with solder or roofing fabric). And the field shingles on the roof lie over the last dormer ridge cap.  □

*Todd A. Smith is a roofing contractor in Verona, New Jersey. For more on asphalt-roofing techniques, contact the Asphalt Roofing Manufacturers Association (6288 Montrose Rd., Rockville, Md. 20852) for a copy of their* Residential Asphalt Roofing Manual, *$10.*

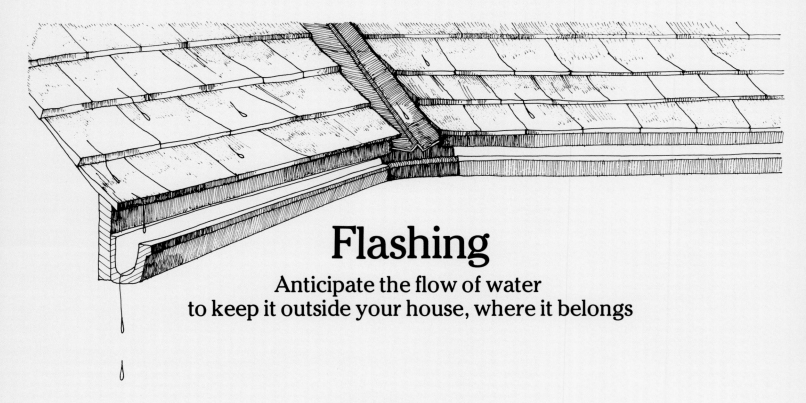

# Flashing

## Anticipate the flow of water to keep it outside your house, where it belongs

### by Bob Syvanen

Flashing is that ounce of prevention worth a pound of cure. I can't begin to count the repair jobs I've done because of faulty flashing. All the mystery was taken out of flashing for me when an old-timer advised that I take a ride on a drop of water as it runs down the roof or sidewall. In so doing, you quickly realize where that drop wants to go, and you understand how to change its direction. Anything that stops the flow will do it. A gutter is a good example. A strong wind will also cause a change in direction, and leaks often result. Other problems oc-cur at junctures of dissimilar materials, where, for example, a chimney meets a roof, or an aluminum window frame joins a sidewall. Any vertical crack is also an open invitation to water—door and window jambs, casings and corner boards.

Flashing used in these and other locations will keep water out of your house. The material used for flashing can be copper, lead, alumi-num, galvanized steel, plastic or paper. What material is used where depends on cost, how severe the condition is, and how long it should last. Copper is the best and the most expensive flashing material. It is strong, long-lasting and can be easily shaped. Lead is right up there in quality and cost; it can be shaped quite easily, but you have to be careful because it punctures and tears. Galvanized (zinc-plated) steel comes next, followed by aluminum, plastic and paper. Unformed metal flashing comes in 10-ft. lengths, in sheets or rolls of various widths and gauges.

*Bob Syvanen is consulting editor to* Fine Homebuilding *magazine.*

---

### Make your own

Although sheet-metal shops are equipped with power shears and brakes that produce crisp lines in a hurry, it is often cheaper and quicker to make and fit your own flashings. This will cut down on travel and turn-around time. The one exception is galvanized steel flashings. Because they are so stiff, it is usually better to purchase them preformed.

Most flashings require a 90° fold (1). The easiest way to get one is to bend the flashing material over a sharp wooden corner. Use a 2x6 with its eased edge ripped off. Then set up a simple jig (2), with nails set along a guide line to position the metal quickly for bending. For narrow bending, a pair of wide-jawed, lock-grip pliers is useful (3). Thin-gauge metal flashing can be cut easily with a sharp utility knife. For heavier metals use tin snips (4) or aviation snips (5) for even easier cutting.

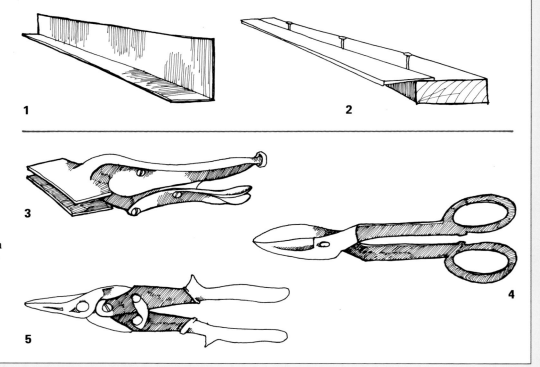

From *Fine Homebuilding* (June 1982) 9:46-50

## Doors and windows

I like to flash window sills with paper, but I use metal on door sills **(6)**, because doors have moisture problems with snow, rain, wet leaves and the like. With window sills, tuck the flashing up into the groove that receives the siding **(7)**. For door and window jambs, flashing paper or 15-lb. felt, 6 in. or more in width, is stapled around the frame **(8)**. The siding felt is tucked underneath this flashing. Make sure that these paper splines run over the sill flashing, which overlays the siding paper. For shingle siding, the sill flashing and the bottom of the jamb spline should overlap the tops of a shingle course, and then be covered by the two remaining courses of shingles underneath the window sill **(9)**. This way the felt won't show through.

Window and door headers must also be flashed, and for this, I think thin metal is best. Nail it in place, keeping the nails high, and continue with the siding. Let the siding felt lap over this head flashing **(10)**. After the siding is on, the ½-in. overhang can be bent over the head casing, using a 2x4 block about 1 ft. long **(11)**. Finish it off by holding the block against the flashing and beating it with a hammer to make a nice flat surface **(12)**.

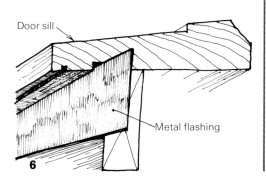

Door sill
Metal flashing

**6**

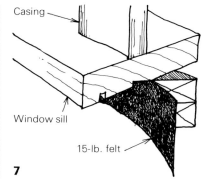

Casing
Window sill
15-lb. felt

**7**

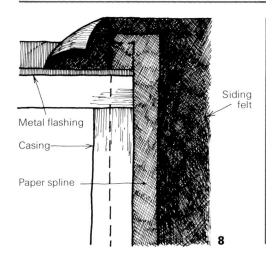

Metal flashing
Casing
Paper spline
Siding felt

**8**

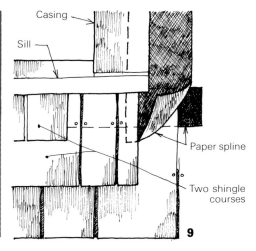

Casing
Sill
Paper spline
Two shingle courses

**9**

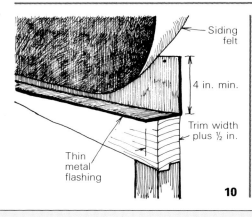

Siding felt
4 in. min.
Trim width plus ½ in.
Thin metal flashing

**10**

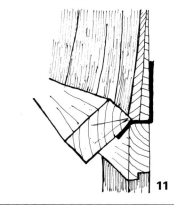

**11**

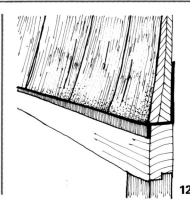

**12**

## Gutters

Another kind of flashing that is made most easily on the job protects wood gutters. Gutter ends that butt against a rake board must be flashed with an oversized piece of lead **(13)**. The flashing can be coaxed into place by gentle tapping with the end of a rubber-handled hammer. The final trimming can be done after the flashing is formed. A wide bed of caulking on the lead edge will seal it, and copper tacks hold it in place. A gutter splice **(14)** is leaded the same way, but there are no compound curves.

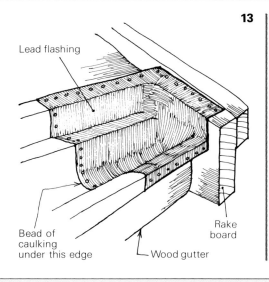

**13**

Lead flashing
Bead of caulking under this edge
Wood gutter
Rake board

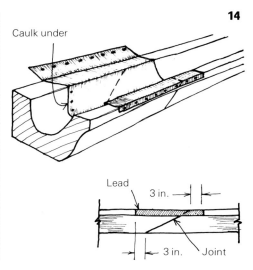

**14**

Caulk under
Lead
3 in.
3 in.
Joint

## Wall and wall-to-roof junctures

Brick-veneer walls require flashing at their bases to keep water away from the wood framing (15). Where a sheet of plywood butts another, the flashing for a horizontal seam is similar to that over a window (16). It is called Z-flashing, and is typically preformed aluminum or galvanized metal in 10-ft. lengths.

Flashing an attached shed roof is hard to do right, and a prevailing wind makes it even tougher (17). Flashing at the peak of a freestanding shed is quite simple (18). One of the toughest conditions to flash is where a pitched roof butts into a wall (19),

but with careful work, water can be kept out. Where a shed roof tucks into an inside corner, the answer is a copper-flashed, tapered cant strip (20).

Step-flash where a roof meets a sidewall (21). Each roof shingle course requires a separate flashing shingle underneath, so I take a roll of galvanized steel, aluminum or copper and cut enough 8-in. squares to do the job. Fold each one 90° over a wood block, and they are ready to nail up. The shingle exposure is typically 5 in., so an 8-in. step-flashing shingle will give you a 3-in. overlap. To keep the shingles in place, nail their upper corners only.

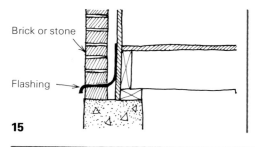

Brick or stone

Flashing

**15**

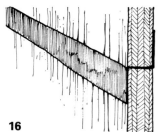

**16**

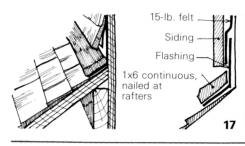

15-lb. felt

Siding

Flashing

1x6 continuous, nailed at rafters

**17**

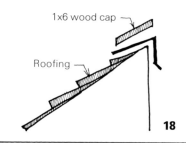

1x6 wood cap

Roofing

**18**

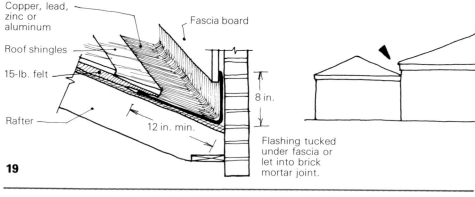

Copper, lead, zinc or aluminum

Fascia board

Roof shingles

15-lb. felt

Rafter

8 in.

12 in. min.

Flashing tucked under fascia or let into brick mortar joint.

**19**

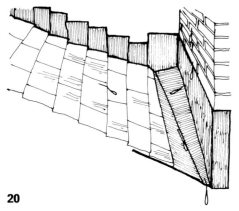

**20**

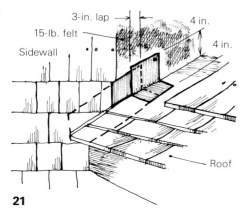

3-in. lap

15-lb. felt

4 in.

Sidewall

4 in.

Roof

**21**

## Chimneys

I think chimneys cause almost as much damage as floods. It's a miracle when they don't leak, sitting up there, buffeted by wind and rain, with at least one big hole on top. The hole on top can be protected with a stone cap, but this doesn't always look good. The very best flashing is thru-pan flashing, and copper is the best material (22). The pan redirects water that gets in at the chimney top. The roof shingles are base-flashed and step-flashed. The brick is cap-flashed over the tops of the base and step flashing (23).

Another common system is to incorporate base flashing and cap flashing with each piece of the step flashing.

Another thru-pan system uses one large lead sheet with a hole cut for the flue, and carefully formed and folded to lie flat over the masonry and on the roof (24).

For chimneys less than 2 ft. wide, a simple flashing bent to the roof pitch will shed rain and snow (25). If the chimney is wider, a cricket is the best solution. It can be a one-piece sheet-metal cricket if it's not too wide (26). A wood-framed cricket can be built for wide chimneys, and handled just like a dormer roof (27).

## Beams

For exterior decks, pressure-treated lumber is best, but if untreated lumber is used, flashing the deck beams will help prevent water damage (28). Intersecting beams should be cap-flashed (29). A beam projecting from a sidewall should be flashed and capped (30), particularly if it is a two-piece beam. Water just loves to sit in that crack between the beams. This solution takes care of the top, but shrinkage at the sides makes caulking the best way to seal this joint. A better solution (31), although not cheap, requires soldering of all joints.

Members fastened to the sidewall are flashed the same way as windows (32). A post through a flat roof is no problem if it is flashed (33). The tops of untreated posts, if left flat, will suck up water like a sponge. Sloping the top will help. A better way is to cap-flash the top. This used to be common practice.

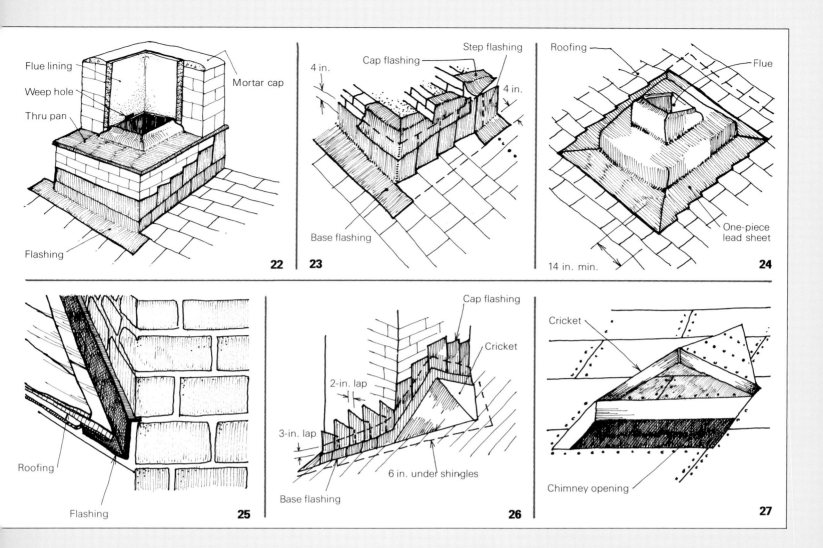

**22**

Flue lining
Weep hole
Thru pan
Flashing
Mortar cap

**23**

4 in.
Cap flashing
Step flashing
4 in.
Base flashing

**24**

Roofing
Flue
One-piece lead sheet
14 in. min.

**25**

Roofing
Flashing

**26**

Cap flashing
Cricket
2-in. lap
3-in. lap
6 in. under shingles
Base flashing

**27**

Cricket
Chimney opening

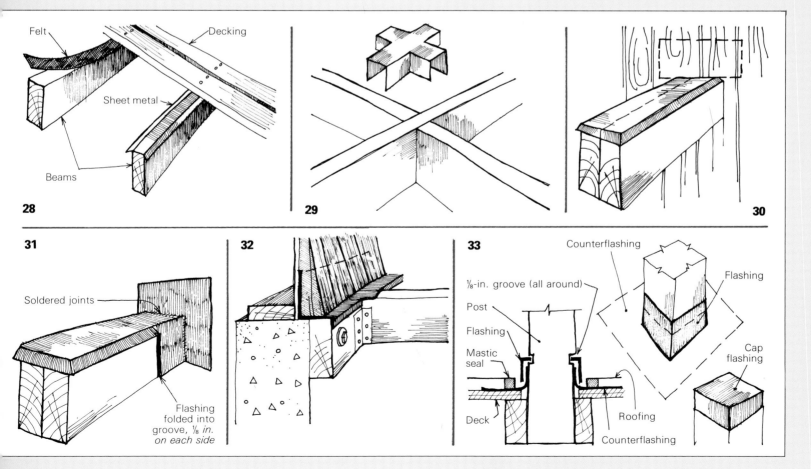

**28**

Felt
Decking
Sheet metal
Beams

**29**

**30**

**31**

Soldered joints
Flashing folded into groove, ⅛ in. on each side

**32**

**33**

Counterflashing
⅛-in. groove (all around)
Post
Flashing
Mastic seal
Deck
Flashing
Roofing
Counterflashing
Cap flashing

## Roofs

Since the neoprene sleeve came on the scene, roof stacks have become easy to weatherproof. You just select the right-sized flange, slide it on the stack, and shingle around it **(34)**. If the roof is wood shingled, check for splits that might leak. If there are any cracks that line up, slip a 2-in. by 8-in. flashing strip, sometimes called a tin shingle, under the split **(35)**. I consider this part of the job whenever a roof is shingled.

Another tough area to flash is a valley. There are basically two kinds: open valleys and closed valleys **(36)**. I like the look of the closed valley, but it is more work. The closed valley starts with a base of 30-lb. felt its full length. The shingles, whether wood or asphalt, are laid to a chalkline snapped up the middle of the felt. The flashings are 12-in. squares of zinc, aluminum or copper, folded diagonally to fit the valley **(37)**. The first piece is a rough half piece. Each shingle course gets one piece held in place with a nail in one outer corner. The shingles are then angle-cut for a nice tight fit.

The open valley is just what it says: open. The valley material can be 90-lb. roofing felt, or copper or galvanized steel. Ninety-lb. roofing felt makes a fine valley if it's installed in two layers **(38)**. Always watch for nails falling into the valley as you work, and keep it clean. The valley should be 5 in. or 6 in. wide **(39)**. It looks better if the valley is the same width on top and bottom, but some recommend widening the bottom to take care of the increase of water. Take your pick, but be sure to cut off the upper inside corner of the asphalt shingles. If the tip is left on, it will catch water, which will ride down the upper edge of the shingle course until it finds a way out, often 15 or 20 ft. away.

Metal valleys are extremely durable, but more expensive. They can be flat in the center, or crimped **(40)**. Some have dams for severe runoff conditions **(41)**. The valley can be nailed at the edges to keep it in place, or you can use clips every 24 in. o.c. **(42)**.

A caution about mixing metals. Galvanic action can be a problem, so don't combine metals from the opposite ends of the following list: aluminum, zinc, tin, lead, brass, copper, bronze. The safest way to proceed is to use clips or nails of the same metal as the flashing.

There is no need for guesswork in flashing. A little thinking goes a long way toward a good job. Just ride that drop of water through the area to be flashed, and you'll know what to do. □

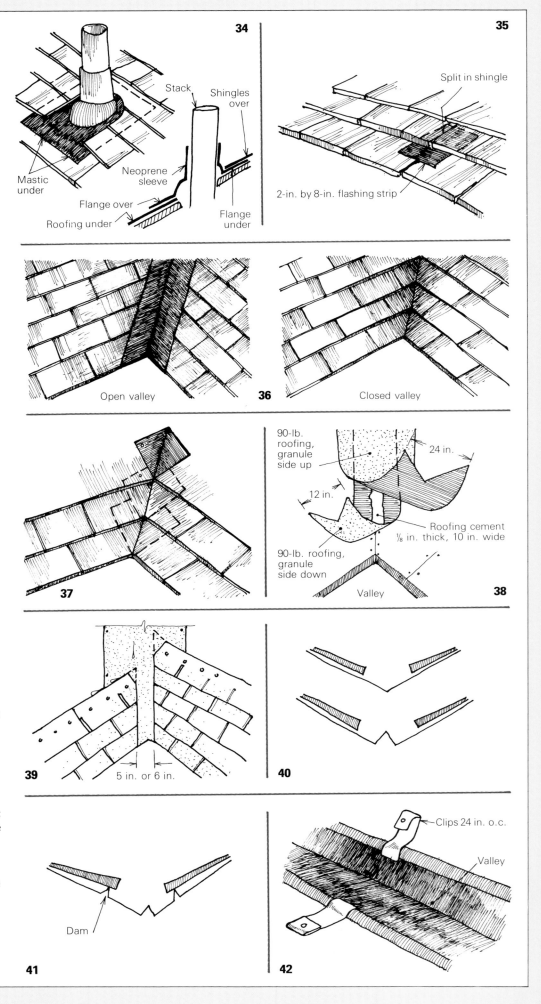

34 Stack  Shingles over  Mastic under  Neoprene sleeve  Flange over  Roofing under  Flange under

35 Split in shingle  2-in. by 8-in. flashing strip

36 Open valley  Closed valley

37

38 90-lb. roofing, granule side up  24 in.  12 in.  Roofing cement ⅛ in. thick, 10 in. wide  90-lb. roofing, granule side down  Valley

39 5 in. or 6 in.

40

41 Dam

42 Clips 24 in. o.c.  Valley

Building and flashing a cricket are the same whether for a masonry chimney or for a wooden chimney structure like the one above, which houses a pair of woodstove chimney pipes. Other metals will work for the flashing, but copper is the easiest to solder.

# Chimney Cricket

## Soldered copper flashing for a rooftop watershed

### by Scott McBride

The best place for a chimney to exit a roof is at the ridge. Water will run away from the chimney, not toward it, and flashing becomes a fairly simple matter. Builders in New England did this back in the days before sheet metal was readily available.

When you move a chimney downslope, however, runoff will strike it, often forcing water up under the flashing and down into the house along the bricks. Leaves and other debris can build up behind the chimney and decompose into a moisture-laden compost that rots roof boards and corrodes the flashing. The way to avoid this mess is to build a well-flashed cricket.

A cricket, or saddle, is a miniature gable roof covered with metal that sits between the back of the chimney and the main roof. The intersection of cricket and main roof produces a pair of valleys that divert water, snow and debris around the chimney. Recently I built a small cricket for a wooden chimney structure

that houses a pair of woodstove chimney pipes. Although the wooden chimney structure made attaching the cricket easier, constructing the cricket and flashing it were much the same as they would be for a masonry chimney:

**Carpentry**—The cricket I built was small enough to make with plywood—no rafters—but I'll speak in terms of common and valley rafters in order to relate the geometry to standard roof framing. Since the chimney structure was 2-ft. wide across the back, the run of the cricket common rafter was half of that (12 in.). The main roof of the house had a pitch of 9-in-12, so to make things easy I gave the same pitch to the cricket. The run of the cricket common rafter being 12 in., the rise was 9 in. and the length, or slope, an even 15 in. You can calculate this using the Pythagorean theorem, but it's easier just to measure the diagonal on a framing square.

Each half of the cricket roof surface (bottom drawing, next page) is a right triangle with one leg 12-in. long (the length of the ridge) and the other leg 15 in. long (the length of the common). The hypotenuse connecting these two sides is the valley. To begin construction, I took a piece of ½-in. CDX plywood 12 in. by 15 in. in size and cut it diagonally, giving me both halves of the cricket at the same time.

To make the pieces fit properly, I beveled the ridge edges at 37°, which is the equivalent of a 9-in-12 common-rafter plumb cut. Then I beveled the edge that runs along the valley at 45°. This was a sharper bevel than was necessary, but the resulting undercut wasn't a problem.

I held the two pieces of plywood in place on the roof and traced the outline of the cricket onto the roof sheathing and the chimney structure. One-half inch below the lines on the chimney, I drew parallel lines marking the top of my nailers. To make these, I cut a

## Flashing layout

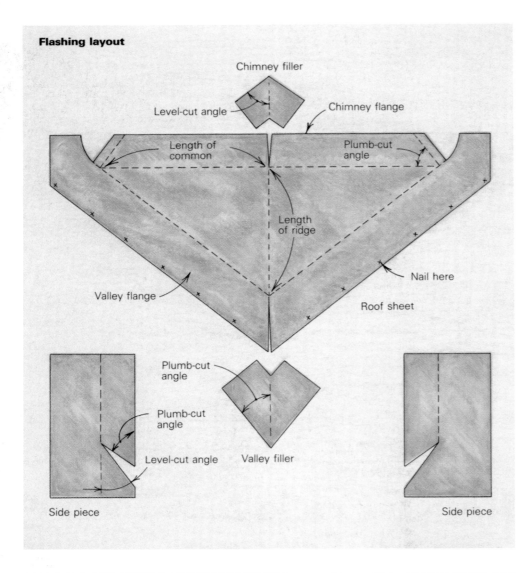

Chimney filler

Level-cut angle

Chimney flange

Length of common

Plumb-cut angle

Length of ridge

Valley flange

Nail here

Roof sheet

Plumb-cut angle

Plumb-cut angle

Level-cut angle

Valley filler

Side piece

Side piece

## Framing and installation

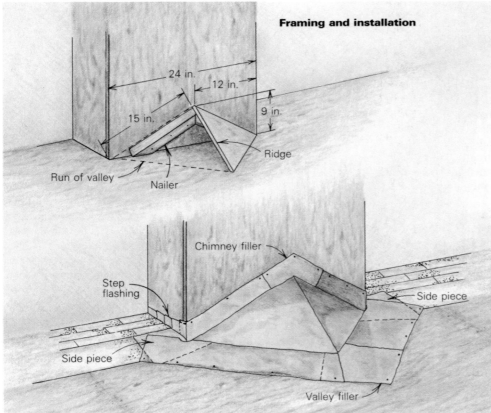

24 in.

12 in.

15 in.

9 in.

Run of valley

Nailer

Ridge

Chimney filler

Step flashing

Side piece

Side piece

Valley filler

pair of 2x4s with a 9-in-12 plumb cut at the top and a 9-in-12 level face cut at the bottom and nailed them in place. I didn't bother with an edge bevel at the bottom since the nailers wouldn't actually have to bear on the roof (drawing, below left). Had this been a masonry chimney, however, these pieces would have been cut as actual rafters. This would have required compound miters on their bottom ends, consisting of a 37° bevel in conjunction with the level face cut. They could then be installed an inch or two away from the brick for heat clearance.

**Flashing**—After nailing down the plywood, I laid out the copper flashing for the cricket. You can use other metals for this, such as aluminum or galvanized steel, but copper is the easiest to solder, and soldering is the best way to seal the joints. Copper comes in sheets and rolls, and is available in different weights and tempers. For this job, I bought 14-oz. hard copper from a local roofing supply. It comes in 3-ft. by 8-ft. sheets, and I paid $45 per sheet.

The flashing assembly consists of one large piece and four smaller ones (top drawing, left). For the large piece, I started with the same dimensions obtained for the plywood, then added 4 in. for a flange against the chimney structure and 8 in. for a flange against the roof. Vertical surfaces, such as the sides of the chimney structure, don't need as much protection as the roof surfaces, which is why I made the chimney flange so much smaller than the valley flange.

Two of the smaller pieces are fillers, covering gaps that open up at the chimney flanges and at the valley flanges when the large, main sheet is folded at the ridge line. The chimney filler was given a notch that was twice the angle formed at the top of each nailer. The valley filler was given a notch that was twice the angle between the valley and the nailer on the cricket's roof surface.

The other two pieces of flashing are side pieces that tie the valley flange, chimney flange and roof sheet together where they meet. The side pieces also turn the corner of the chimney structure and overlap the step flashing coming up along the chimney's side.

To make the side pieces, I started with a rectangular piece of copper and cut a slot in from one edge at 37° (the plumb cut angle). This allowed a flange to be turned up that would butt into the upturned chimney flange. I left tabs on the ends of the chimney flanges to create an overlap with the side pieces.

Most of the bending (top photo, facing page) was done on a portable aluminum J-Brake (Van Mark Products Corp., 24145 Industrial Park Drive, Farmington Hills, Mich. 48024). But after making a series of bends, I could no longer slip the roof sheet into the brake and had to form the ridge crease over the edge of a 4x4 post. I also used hand seamers for some of the smaller bends. These are like pliers with broad flat jaws and are made specifically for working sheet metal.

Drawings: Michael Mandarano

To facilitate assembly and ensure accurate positioning of the pieces while soldering, I built a mock-up of the chimney structure. The mock-up allowed me to work in the shop and keep the heat of soldering away from the building itself—something that's always made me a bit nervous when soldering flashing in place.

**Soldering copper—**The secret of the tin-knocker's art lies not in the soldering itself, but in the fabrication of the parts. Close-fitting joints will suck in the molten solder just like joints in copper plumbing do, which makes sound, neat joints requiring minimal heat. Big gaps, on the other hand (anything over 1/16 in.), will require lots of solder and repeated heatings to seal them. Also, the excess heat causes the metal to warp, which drives the mating pieces even farther apart. The result is a thick, sloppy bead that may not be watertight. Once this happens, it's too late to go back and refit the copper, so the job has to be done right the first time.

After cutting out and bending the pieces, I cleaned the contact surfaces with steel wool (emory cloth works, too). New copper doesn't need much of this, but old tarnished copper requires mountains of elbow grease to get clean. Next, I carefully applied flux only to the areas where I wanted the solder to stick. Limiting the application of the flux in this way helps contain the molten solder once it starts to flow and enables me to peel hardened solder from those areas left un-fluxed. I brushed on a liquid-type acid flux, as opposed to the paste flux preferred by plumbers. The liquid leaves less residue and is easier to wash off. I'm very careful not to get it on me—especially not in my eyes.

Before soldering each joint, I arranged the pieces so that the joint to be soldered was level. This reduced the tendency of the liquified solder to run away from the joint. I then clamped the pieces in place, keeping the clamps far enough away from the joints so they wouldn't act as heat sinks. Locking pliers, especially the deep-reach kind, are my favorite clamps for this. To keep the joints tight, I had a helper press down on either side of the flame with a pair of screwdrivers and move along with me as I soldered. This type of "moving clamp" is a great help.

The solder used by roofers is half lead and half tin. It comes in 1-lb. bars about 10-in. long. Traditionally, roofers melt it with a copper soldering iron heated in a charcoal-fired brazier. I've never had good results with this tool, however, so I use a propane torch. The drawback to the hardware-store-variety torch is that most are cheaply made—some of the ones I've used barely have outlasted the fuel canisters they were mounted on. An alternative is the heavy-duty air-acetylene torches commonly used by plumbers.

After adjusting the torch to burn with a sharp blue flame about 1½-in. long, I use a circular motion and play the tip of the flame over a broad area to warm up the copper a

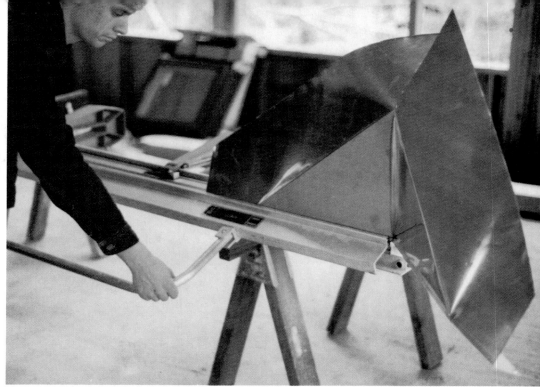

Most of the folds were made on the metal brake shown above. Eventually the configuration of bends prevented the roof sheet from slipping into the brake, so the final folds were handmade.

bit. As I do this, I hold the bar of solder close to the flame so that it will also catch some of the heat. When the flux starts to sizzle, I put the tip of the flame down onto the work at the beginning of the joint. Then I press the tip of the solder bar against the work about ½ in. from the end of the flame and let the heat flow through the copper to melt the bar. I move the flame along the joint, followed by the solder bar (photo right). When you do this, you have to move at an even, moderate pace. If you move the flame too fast, the solder won't flow, but you can't linger either or you'll warp the metal with too much heat.

Textbooks on soldering will tell you that the heat should always be conducted through the work as I've just described, and that you should never melt the solder directly with the flame, dripping blobs of molten solder onto the joint. Well, that's fine if you've got a perfect joint lying flat on the bench.

But sometimes—especially if you have to work up on the roof, with your pieces at an angle, or if the pieces don't fit real tight—there won't be enough capillary action to keep the liquified solder from running out of the joint as fast as it runs in. The only practical way to deal with the situation is to pull the heat back a little so that the copper is hot—but not *that* hot. If you then drip molten solder onto a slightly cooled joint, it'll stick. It might not be as strong as a properly soldered joint, but at least it won't leak. And it's still better than a dollop of roofing tar.

After soldering, you have to clean off any remaining flux with water and a stiff brush in order to prevent corrosion.

**Installation—**Before cloaking the wooden cricket in its copper cape, I shingled up the sides of the chimney structure, inserting step flashing as I went. When I got up to the back

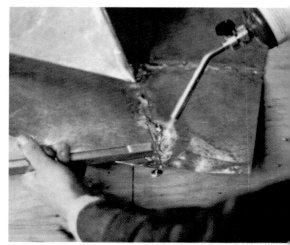

McBride solders the flashing in his shop, instead of on the roof. This way he can position the flashing so that the seams are level when he solders them, which keeps the molten solder from running out of the joint.

of the chimney, I slipped the sidepieces of the cricket flashing down over the uppermost shingle course and the top piece of step flashing. I anchored the cricket flashing with roofing nails sunk along the outside edges of the valley flange and chimney flange only (photo on p. 111). Then I continued with the roofing, bringing my shingles up to the centerline of the valley, but keeping all roofing nails off the flashing itself.

Since this was a wooden chimney structure, the vertical fins of the step flashing and the chimney flange were eventually covered by siding. Had this been a masonry chimney, a separate cap flashing, tuckpointed into the brickwork, would have been used as coverage. □

Scott McBride is a carpenter in Irvington, N. Y. and contributing editor to Fine Homebuilding. Photos are by the author.

***Extending the rake.*** **For a clean look, the top edge of the rake's crown molding extends down past the edge of the roof, where it nearly meets the corner of the gutter. The crown is protected by a tongue of asphalt roofing over a flap of aluminum flashing.**

# Rethinking the Cornice Return

## Standard ogee gutters can mimic this classical detail

### by Scott McBride

**C**onstruction of traditional cornice returns (top photo, p. 116) is a lost art. Like many of the finer points of residential carpentry, returns were cast aside after WWII in the rush toward mass-produced housing. They were replaced by the "pork chop," or ear board, a no-nonsense way of resolving the transition from rake board to eaves fascia. The pork chop is an unadorned triangular piece of pine tacked up flush with the rake board (bottom right photo, p. 116).

A generation of carpenters matured in the 1960s and 1970s with no other return in their repertoire. Then the post-modern movement came along, and suddenly the classic cornice return had to be reinvented. Unfortunately, there weren't many old-timers left who remembered how to build one (for Bob Syvanen's treatment of cornice returns, see the article on pp. 118-121). The task was complicated further because guttering practices had changed in the interim.

In the 18th and 19th centuries, gutters usually were built into the roof and lined with sheet metal. With the gutter concealed, the cornice was free to return neatly onto the gable. Unfortunately, when concealed gutters leak, the water drips into the cornice or, worse, into the wall. In the late 1800s, builders started switching to metal half-round suspended gutters. But these gutters not only look bland, they also throw the crown molding—the crowning glory of the cornice—into

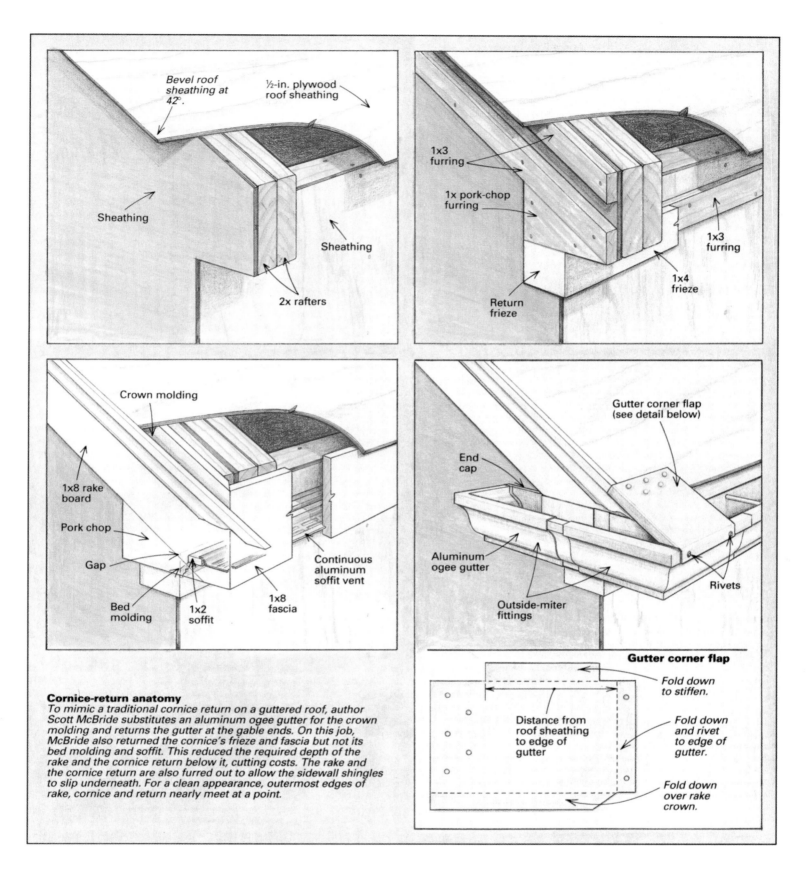

**Bevel roof sheathing at 42°.**

½-in. plywood roof sheathing

Sheathing

Sheathing

2x rafters

1x3 furring

1x pork-chop furring

1x3 furring

1x4 frieze

Return frieze

Crown molding

1x8 rake board

Pork chop

Gap

Bed molding

1x2 soffit

1x8 fascia

Continuous aluminum soffit vent

Gutter corner flap (see detail below)

End cap

Aluminum ogee gutter

Outside-miter fittings

Rivets

**Gutter corner flap**

Distance from roof sheathing to edge of gutter

Fold down to stiffen.

Fold down and rivet to edge of gutter.

Fold down over rake crown.

**Cornice-return anatomy**
To mimic a traditional cornice return on a guttered roof, author Scott McBride substitutes an aluminum ogee gutter for the crown molding and returns the gutter at the gable ends. On this job, McBride also returned the cornice's frieze and fascia but not its bed molding and soffit. This reduced the required depth of the rake and the cornice return below it, cutting costs. The rake and the cornice return are also furred out to allow the sidewall shingles to slip underneath. For a clean appearance, outermost edges of rake, cornice and return nearly meet at a point.

shadow (bottom left photo, p. 116). The metal ogee gutter, or K gutter as it's sometimes called, seems to have appeared in the 1930s. It's an ingenious idea; combine the elegant look of a traditional crown molding with the practicality of a surface-mounted gutter. Given that the strong, easy-to-install ogee gutter is here to stay, I've long sought a way to incorporate it gracefully into a traditional cornice return. After several false starts, I think I've succeeded.

**Returning the gutter—**My first attempt to integrate the ogee gutter into a traditional return consisted of simply mitering the end of the gutter. It wrapped around the corner over the pork chop and then terminated at an end cap. This looked okay, but it sure didn't sing like the originals.

Then I realized that wooden cornice returns don't just stop once they turn the corner—they turn *another* corner and dead-end at the gable wall. Okay, so I connected a second outside

miter fitting to the first one, with a short length of gutter in between. One leg of this second fitting had to be cut off, leaving just enough room for an end cap to be riveted on. Visually, this helped quite a bit, but there was still something wrong. After studying a lot of old-style cornice returns, it hit me: The outermost edges of the rake, the cornice and the cornice return should all meet at one point, or at least appear that way from the ground. The eye can't help but follow these

lines, and if they converge neatly, a little visual ping goes off.

One problem with my gutter return was that the lip of the eaves gutter sat above the projected roof plane. When viewed from the gable end, the gutter jutted above the rake line. Another problem was the way the return looked from the front. It stuck out 4 in. past the flat rake board like a cow's ear. No wonder it looked clunky.

**Working out the details—**On a recent job in Rye, New York, I was determined to get my cornice act down. I started drawing the cornice details before the second-floor wall framing was up.

I needed three drawings: a vertical section through the eaves, a section through the rake cut perpendicular to the roof and a horizontal section of the eaves-end cornice return, which is a sort of plan view. My goal was to make the total projection of the eaves, including the gutter, equal to the projection of both the cornice return and the rake. This would allow all three edges to come to a point.

I started with the eaves section, drawing a 2x6 wall with a 2x10 rafter on top of it at the given pitch. Then I sliced thin sections of the different trim elements to use as templates and juggled them around to find a workable configuration. I knew that extending the rake would be the most difficult part; to keep eaves and rake projection roughly equal, I designed the eaves with a minimum overhang.

To keep the eaves overhang as shallow as possible, I called for a narrow 1x2 soffit and slipped the fin of the continuous aluminum soffit vent into a groove plowed in the back of the fascia (drawing, p. 115). The fascia itself would be 7½ in. wide, allowing me to use 1x8 stock for it without ripping. The wider-than-usual fascia was also part of my design strategy; it would allow me to align the outside lip of the gutter with the plane of the roof so that the top edge of the gutter would meet the line of the rake at the corner.

Returning all of the cornice elements onto the gable, minimal as they were, would produce an 11¼-in. overhang for the cornice return. To align the rake with the corner, I would then have to frame an 11¼-in. rake overhang—more work than I had contracted for. I compromised by returning only the fascia and the frieze, in addition to the gutter (photo, p. 114). This would keep the lip of the gutter return close to the house so that it would nearly align with the lip of a 5½-in. crown molding mounted on a furred-out rake board.

A section drawing of the rake helped me figure out the overhang needed on the roof sheathing to support the top of the rake crown molding. The drawing also made it easy to determine the cutting bevel along the edge of the sheathing, which depends on the particular crown used. During installation we let the plywood run long, then snapped a chalkline at the correct distance and cut it with a circular saw set at 42°.

**Corner flaps—**To bring the rake crown molding down to the corner of the gutter without exposing the molding to the weather, I took my cue from the built-in gutters I've seen. They typically have a narrow tongue of roofing at each end that

*Cornice evolution.* Traditional cornice returns often featured a soffit and a full complement of cornice moldings capped by a flashed water table that prevented water damage. In the example above, the top edges of the rake, the cornice and the return meet at a point for an uncluttered look. In the late 1800s, half-round metal gutters began to supplant less-dependable built-in gutters. Bland in appearance, the half-round gutters obscured crown moldings (photo below). Nowadays, standard ogee gutter routinely substitutes for crown molding and simply dead-ends at the corners of the roof. The pork chop, an unadorned 1x triangle that's installed flush with the rake board (photo right), has virtually replaced the traditional cornice return.

Photos this page: Bruce Greenlaw

extends down to cover the lower end of the rake crown molding. The built-in gutter dead-ends at the inboard edge of the tongue. Around the corner, a flashed water table caps the cornice return's crown molding (the water table is simply a piece of wood that's sloped away from the house to shed water). The rake crown dies into the top of the water table.

I mimicked this arrangement. Instead of extending the roof sheathing to support the tongue of asphalt roofing, though, I installed a simple flap made from white aluminum coil stock (coil stock is trade jargon for prefinished aluminum on a roll). The edges of the flap were turned down with hand seamers to stiffen them, and a fin in front laps over the lip of the gutter. But as it turned out, the gutter used by my gutter contractor is wider than the sample I had obtained from a lumberyard. As a result the flaps fell short, but they still get enough support from the extended rake crown. The tongues of the asphalt roofing were glued to the flaps with roofing mastic.

**Splitless nails**—Because of all my planning, installation of the cornices went like clockwork. I set up a cutting bench with a Hitachi 15-in. miter saw recessed into the top. This machine swings to 58°, so I could even cut the acute miter at the bottom of the rake crown molding.

The job's general contractor, Frank DiGiacomo, showed me an exterior trim nail that I had not seen before. Called a "splitless," the nail is made by the W. H. Maze Co. (100 Church St., Peru, Ill. 61354; 815-223-8290). A splitless nail is a cross between a finish nail and a box nail. The small, flat head makes it easier to drive than a finish nail and resists trim warpage. The narrow diameter of the shank is less likely to split wood, which is how the nail got its name. The splitless nail has a hot-dipped zinc coating for corrosion resistance and is available in 6d, 8d and 10d sizes. I used these nails exclusively for assembling the cornice.

**A variation at the entry**—On the entry porch of the Rye project, where the return would be more visible, I decided to frame a full rake overhang by notching 2x4 lookouts into the end rafters (drawing right). This full overhang allowed me to maintain a constant soffit width for the eaves and the cornice return, the latter of which would line up with the rake crown (photo right). Because there is no gable wall for the cornices to return to, the returns make one extra 90° bend and dead-end into the porch ceiling.

I had intended to use aluminum gutters in place of crown molding for the cornice returns as I had done on the higher gables. But once the gutter was installed, it looked way too wide. I replaced it with a 4½-in. crown molding capped by a flashed water table. This looks good but leaves the rake crown sticking a bit proud of the cornice. It isn't too noticeable, though. When I lived in New York we used to say, "Fugetaboutit." Here in Virginia, they say, "It is what it is."  □

*Scott McBride is contributing editor of* Fine Homebuilding *and lives in Sperryville, Va. Photos by the author except where noted.*

**Return of the cornice.** Because the cornice return on this entry porch is so prominent, the author framed a full rake overhang, which maintains a constant soffit width for the eaves and the cornice return. The drawing below shows how the rake overhang is built.

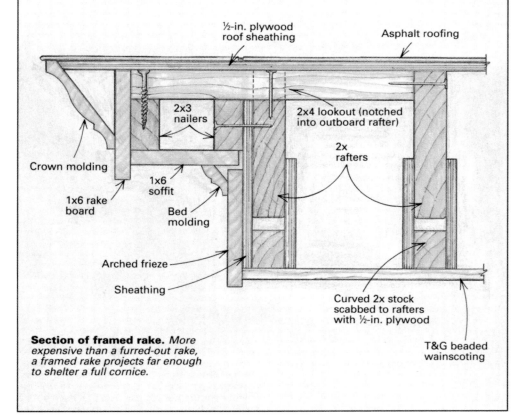

½-in. plywood roof sheathing

Asphalt roofing

2x3 nailers

2x4 lookout (notched into outboard rafter)

2x rafters

Crown molding

1x6 soffit

1x6 rake board

Bed molding

Arched frieze

Sheathing

Curved 2x stock scabbed to rafters with ½-in. plywood

T&G beaded wainscoting

**Section of framed rake.** *More expensive than a furred-out rake, a framed rake projects far enough to shelter a full cornice.*

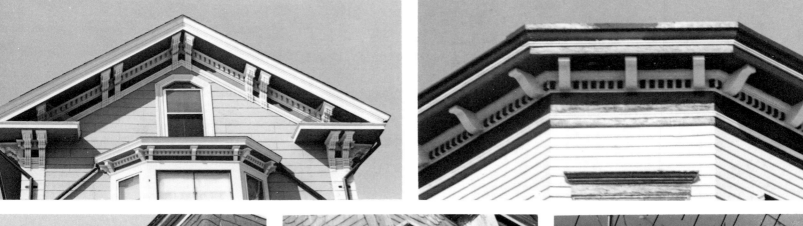

# Cornice Construction

## Building the return is the tough part

by Bob Syvanen

All gable-roofed houses need to have a cornice of some sort. Functionally, the cornice fills the voids between roof and sidewall. It extends from the shingles to the frieze that covers the top edge of the siding. There are two basic kinds of cornices: one includes a gutter (drawing **1A**, below); the other is a simple cornice molding (**1B**). Traditionally, each calls for a different method of roof framing. For the gutter, rafter tails are combination-cut (plumb and level) and bear on the top plates. For a cornice without the gutter, a double plate is nailed across the tops of the joists to support the rafters. There are a vast number of possibilities in cornice construction and detailing (photos, facing page), but they are all variations on the basic, step-by-step sequences shown and discussed here.

The cornice return at the juncture of eave and gable is the most noticeable and most intricate part of this architectural detail. Sometimes called a boxed return, it adds a delicate touch to the large, repetitive detail of clapboard or shingle siding.

Building the cornice itself is simple; building its return is a bit more challenging. Let's start with the eave corner, assuming that you've already framed and sheathed it. The rafter ends and ceiling joists should still be exposed, and the first step is to nail up the fascia board. For the guttered cornice (**2A**), the fascia will go against the rafter tails; otherwise, nail it to the ends of the ceiling joists (**2B**). For the fascia, I use #2 pine boards ripped to width on a table saw. The top edge of the fascia usually needs to be beveled for a tight fit.

At the corner of the house, the fascia meets the ear board—the broad, flat backing board at the gable base that holds the cornice return. The joint between the fascia and the ear board has to be mitered, and unless the fit is

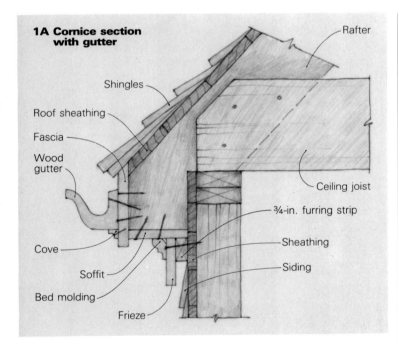

**1A Cornice section with gutter**

Rafter · Shingles · Roof sheathing · Fascia · Wood gutter · Cove · Soffit · Bed molding · Frieze · Ceiling joist · ¾-in. furring strip · Sheathing · Siding

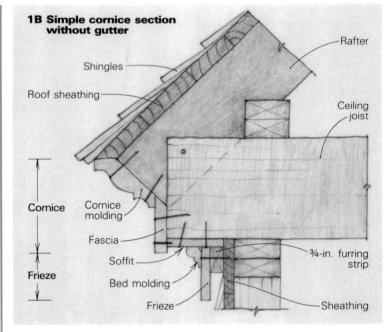

**1B Simple cornice section without gutter**

Rafter · Shingles · Roof sheathing · Ceiling joist · Cornice · Cornice molding · Fascia · Soffit · Frieze · Bed molding · ¾-in. furring strip · Frieze · Sheathing

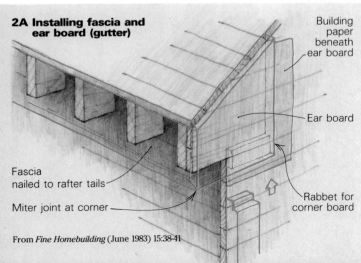

**2A Installing fascia and ear board (gutter)**

Building paper beneath ear board · Ear board · Fascia nailed to rafter tails · Miter joint at corner · Rabbet for corner board

From *Fine Homebuilding* (June 1983) 15:38-41

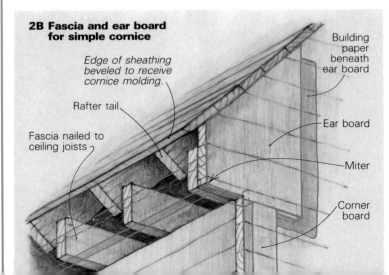

**2B Fascia and ear board for simple cornice**

Edge of sheathing beveled to receive cornice molding. · Rafter tail · Fascia nailed to ceiling joists · Building paper beneath ear board · Ear board · Miter · Corner board

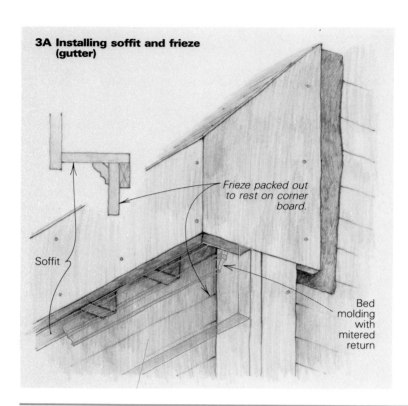

**3A Installing soffit and frieze (gutter)**

Frieze packed out to rest on corner board.

Soffit

Bed molding with mitered return

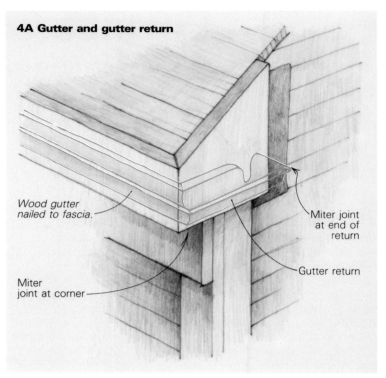

**4A Gutter and gutter return**

Wood gutter nailed to fascia.

Miter joint at corner

Miter joint at end of return

Gutter return

tight, it's a good idea to use a weatherproof adhesive caulk. Phenoseal (made by Gloucester Co., Box 428, Franklin, Mass. 02038) and other marine adhesives work well in this type of joint. I also staple a piece of building paper behind the ear board.

If you want the ear board to be flush with the corner board, which is installed before the siding is nailed up, the ear board needs to be rabbeted at its bottom edge so that the siding and a tongue at the top of the corner board can be tucked up underneath it. The ear board should also cover the gable edge of the roof sheathing if you're not planning to use

rake molding, which is shown in the inset drawing **6B**.

Now install the soffit, frieze and bed molding, in that order. I cut the soffit and frieze boards from pine or fir, and fur out the frieze a full ¾ in. so that the siding will fit underneath it. Cut the frieze to extend over the corner board, as shown in **3A** and **3B**. You can use something fancier than bed molding to cover the joint between soffit and frieze, or simply leave this juncture plain.

**The return**—The next piece to go on is the cornice (crown) molding (**4B**), or the wood

gutter that replaces it on a guttered cornice (**4A**). Here again, the corner is mitered to make the return. In fitting the return, I like to tack a temporary guide strip to the ear board, on a level line where the bottom of the cornice molding or gutter will fit. This makes test-fitting, trimming, and retesting the miter a bit faster.

A good tight joint here is important. Corners are seldom perfectly square, and you can either adjust the cut in the miter box or just keep fitting and trimming with a chisel or block plane. The small triangular piece at the end of the return is delicate, and I usually cut

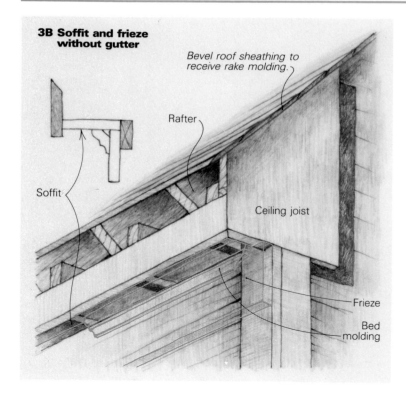

**3B Soffit and frieze without gutter**

Bevel roof sheathing to receive rake molding.

Rafter

Ceiling joist

Soffit

Frieze

Bed molding

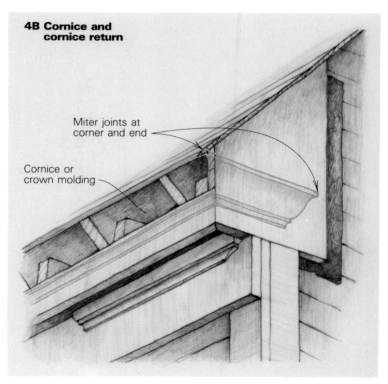

**4B Cornice and cornice return**

Miter joints at corner and end

Cornice or crown molding

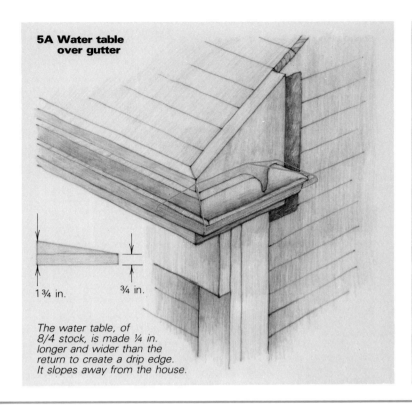

**5A Water table over gutter**

1 3⁄4 in.    3⁄4 in.

*The water table, of 8/4 stock, is made 1⁄4 in. longer and wider than the return to create a drip edge. It slopes away from the house.*

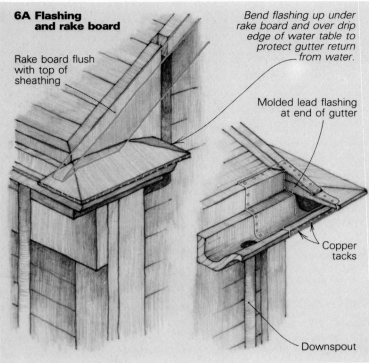

**6A Flashing and rake board**

Rake board flush with top of sheathing

*Bend flashing up under rake board and over drip edge of water table to protect gutter return from water.*

Molded lead flashing at end of gutter

Copper tacks

Downspout

and fit it on the ground after the corner miter is done.

What you have now is the mitered return with an opening on top. This opening must be covered with a water table (**5A**, **5B**), which is a piece of 8/4 pine, beveled away from the side of the house to shed water. You can cut a single bevel, or bevel the board downward on all open sides for a fancier look. Size the board a little longer and wider than the return to create a drip edge. The water table for the simple cornice return has to be trimmed to fit beneath the roof sheathing (inset, **5B**).

Once you've nailed the water table in place, flash it before installing the rake board along the corner between the roof and the gable wall. The flashing has to extend up beneath the rake board, so if you're working on an old house, pry the rake board up, tuck the flashing underneath it, and then renail the rake. If you are going to use rake molding (**6B**), nail the rake board to the gable end, furring it out away from the sheathing at least 3⁄4 in. so that the siding can be fitted underneath. To provide good bearing for the rake molding, bevel the gable-end sheathing. If you don't use a rake molding, leave the sheathing square-cut so that the rake board and its furring strip

(**6A**) can be nailed to it. In either system, the rake board and molding die onto the water table, and must be scribed accordingly.

If you're building a guttered return, there's one final, important step: sealing off the corner. Water that gets into the return side of the gutter will soon rot the wood. Lead is the best flashing material because it can be formed to the gutter contours more readily than copper or zinc. Use the end of your hammer handle to form the lead flashing so that it covers the corner and fits over the water table, rake board, and roof sheathing. Then nail it down with copper tacks in a bed of caulking.  □

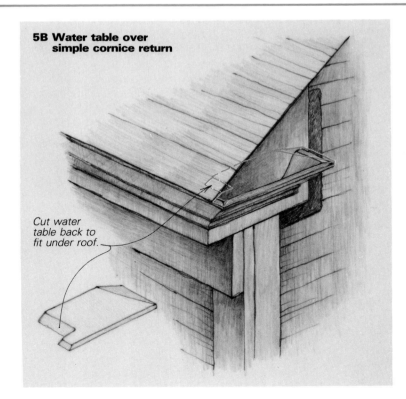

**5B Water table over simple cornice return**

Cut water table back to fit under roof.

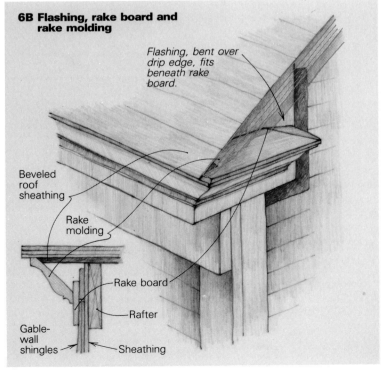

**6B Flashing, rake board and rake molding**

*Flashing, bent over drip edge, fits beneath rake board.*

Beveled roof sheathing

Rake molding

Rake board

Rafter

Gable-wall shingles

Sheathing

# Replacing Rain Gutters

## A professional's advice on installing aluminum and galvanized-steel gutters

by Les Williams

When I installed my first set of gutters a dozen years ago, my boss started me out with these bits of information: "Water always takes the path of least resistance, never return to the truck empty-handed and payday is on Friday." Now that I've got my own sheet-metal and gutter-installation business, I know from experience that two of those statements are always right. Paydays, unfortunately, aren't always on Fridays when you work for yourself.

I work around the San Francisco Bay Area, where nine out of ten of the gutter jobs that I do involve replacing worn-out gutters on existing homes. Although replacing gutters presents its own set of challenges, the principles are the same for new construction—replacing gutters is just harder to do.

Clients sometimes ask, "Why do I even need gutters?" For one, in some jurisdictions the code insists on them. But more than that, gutters just make sense in areas that have periods of heavy rainfall. They can prevent water from splashing against the house, causing premature paint failure and wearing out window frames. And they keep heavy runoff from eroding the soil around foundations. As David Helfant pointed out in his article on drainage (*FHB* #50, pp. 82-86), diverting roof runoff away from a foundation can make a big difference in the stability of some kinds of soil. Also, in many cases the gutters serve as trim for an otherwise unfinished-looking eave.

**Styles, sizes and materials**—Gutters are one of those architectural elements that can be put in the spotlight as decoration, or blend into the lines of the roof. Off-the-shelf gutters range from the simplicity of the fascia gutter (top photo), to the more traditional ogee style (bottom photo). A simple half-round gutter is also a stock profile. If your house requires something more distinctive, sheet-metal shops can custom-make gutters with almost any combination of reveals, coves and S-curves that you may need.

Ogee and half-round gutter dimensions commonly range from 4 in. to 6 in., measured across the top from the back of the gutter to the front. Fascia gutters are an exception. They are called 5½-in. or 7½-in., which refers to their depth.

I haven't seen any stock metal gutters that weren't large enough to handle the amounts of rainfall in the Bay Area. The primary con-

A fascia gutter is a simple U-shaped channel with a flat face that tilts outward. Here the author checks a section for positive drainage toward the outlet.

The face of an ogee gutter has an S-curve and a couple of reveals to cast shadow lines. Here the author drills a hole through an outside miter strip prior to pop-riveting the miter strip to the gutter section.

sideration in getting rid of the water is the size and number of downspouts. A rule of thumb for determining downspout requirments is that a 2x3 downspout can take the average rainfall from 600 sq. ft. of roof. Another way to figure it is 1 sq. in. of downspout cross section per 100-sq. ft. of roof. This downspout-to-roof ratio will drain runoff in Dade County, Florida, which receives the most intense rainfall in the continental United States.

Standard gutters are available in aluminum, galvanized steel, wood, copper and plastic. I seldom work with wood and I don't work with plastic gutters. While I like to make and install copper gutters, they are pretty pricey. The cost for the materials for the aluminum and galvanized gutters shown in this article is roughly $1/ft.; copper is five times more expensive.

**Making comparisons**—In this article, I'll talk about the two most common types of gutters that I install: aluminum ogees and galvanized-steel fascia gutters. Both materials have their assets and their drawbacks.

On the plus side for aluminum, the kind I install (Alcoa Building Products, P. O. Box 716, Sidney, Ohio 45365; 800-621-7466) are prepainted and carry a good warranty. They won't rust or corrode, they don't require soldering and they have a slick expansion-joint system. They are large-capacity gutters, and Alcoa can supply different kinds of hangers for varied applications that don't restrict the movement of the gutters, allowing them to expand and contract without buckling, breaking seams or working the fasteners loose.

On the other hand, some clients consider the preformed "miter strips" that join the sections of gutter at 90° corners to be visually obtrusive, and site-formed odd-angled joints are potential weak spots. Alcoa's corrugated downspouts aren't beautiful; the gutters only come in two sizes of ogee and the colors are limited.

Galvanized gutters are available in many styles and sizes, and the site-formed, soldered intersections used to join them at splices and corners are strong and clean-looking. Galvanized gutters are stronger than aluminum, and their round- or rectangular-section downspouts look good.

The downside to galvanized gutters includes a higher installation cost than aluminum because of the soldering involved (I charge about 25% more). Galvanized gutters have to be painted, eventually they rust out and the hanger selection is limited.

**Hangers**—Galvanized and aluminum gutters are secured to the roof in different ways. The aluminum gutters that I install are attached by way of hangers. There are basically two types: roof-mounted and fascia-mounted (drawings facing page). Most roof-mounted hangers are one-piece affairs that are nailed through the sheathing and into the rafters. The flange on the fascia-mounted hanger allows it to be nailed to the side of the building, making this hanger the choice for replacement gutters because it can be installed without disturbing the roofing. The hooks on the ends of the hangers engage the lips on the outside edges of the gutter, supporting its weight. They also hold the front of the gutters in a straight line and help to keep the gutters from being crushed when someone leans a ladder against them.

The two-piece hanger consists of a threaded rod that is installed the same as the one piece hanger, but it has a separate brace, which is attached to the gutter and is adjustable by means of two nuts on either side of the gutter brace. They are typically used with decorative rafter tails. This hanger also allows you to slope the gutters after the hanger is installed.

Alcoa makes three roof-type hangers and three fascia-type hangers. Installed correctly, they all work well. My favorite fascia-type hanger has a turned-down nailing flange making it a little easier to install when a row of shingles is overhanging it. This hanger also allows the gutter to be installed a little closer to the shingles, which I think is more attractive and better for catching heavy runoff.

Galvanized gutters are nailed through the back to the fascia. Then a spacer called a strap-hanger is affixed to the roof that also supports and aligns the front of the gutter. One type of fascia-mount system that should be avoided is the spike-and-ferrule. It consists of a 7-in. spike that is driven through the front of the gutter, then through a spacer tube, through the back of the gutter and into the fascia or the ends of the rafters. If the spikes don't split the rafters right away, the weight of the water and the expansion and contraction of the gutters will be enough to cause the spikes to loosen and eventually fail.

**Getting started**—Task one is to decide where the downspouts should go. They should be in unobtrusive locations, out of traffic patterns. In cold climates, they are best kept away from north walls to reduce the chance of freezing. Next I decide how I am going to lay out the gutters so that I have as few joints and scraps as possible. Most of the gutter I use comes in 20-ft. or 21-ft. lengths. If I have a section that is 22 ft., I will usually install a piece about 17 ft. long and another piece to fill out the run. It isn't a good idea to install a very short piece at the end of a long piece because it is difficult to make the whole thing look straight. I also try not to put a seam right over the front entrance of the house. Common sense is the key.

I don't slope gutters more than ¾ in. in 20 ft. because they start to look out of kilter. The worst situation is when severe settlement pitches the roof in the direction opposite the downspout. When that happens, the downspouts should be moved. I typically start a gutter job at the hardest part of the house—either the highest wall, the most complicated corner or the toughest place to reach.

After I've torn off the old gutters, I work 1-in. by 3-in. metal flashing under the existing roofing. The flashing is made to the pitch of the roof out of 26-ga. galvanized sheet-metal, and it prevents runoff from wicking behind the gutter or along the bottom edge of the roof sheathing.

**Installing aluminum gutters**—Up to now, I could be installing either galvanized or aluminum gutters. The hangers are one detail that set the two apart; I'll talk about the aluminum ones first.

With the flashing in place, I nail a hanger at the high point of the gutter section and another at the low point. If the section is 20 ft. long or less, with end caps on both ends, very little slope will be necessary. I snap a chalkline along the bottom of the hangers to guide the placement of the rest of the hangers. Then I nail on the remaining hangers, one every other rafter (no more than 5 ft. apart). It is often necessary to notch the flashing to accommodate the hanger tops. Alcoa says to use aluminum screw-shank nails for attaching the hangers, but I don't. They bend too easily, so I use 8d hot-dipped galvanized nails. From what I've read and observed, galvanized steel next to aluminum doesn't cause a significant corrosion problem in this kind of application.

Now I measure from corner to corner to see exactly how long a piece of gutter I need. If one end terminates at a gable, I extend the gutter ½ in. past the gable to catch all of the water coming off the roof.

Adjacent gutter sections that join at 90° angles are connected to one another by way of miter strips that overlap the gutter ends (bottom photo, facing page). If a gutter section between two miter strips is too long, it can cause problems. So I subtract ⅛ in. per miter strip from my overall gutter measurement.

**Cutting gutters**—To cut the gutter sections, I use either a good old-fashioned hacksaw with a 32-tooth per in. blade or a Makita 14-in. miter saw with a 120-tooth carbide blade. I only use the miter saw for aluminum or copper gutters, and I use a contoured block of wood inside the gutter to back up the cut.

When cutting the gutter with a hacksaw, I first mark the miter or straight cut on the bottom and back of the gutter. Then I grasp the gutter with my left hand and begin my cut with the hacksaw blade at approximately a 45° angle from the bottom of the gutter (top photo, next page). Using light-to-medium pressure, I follow the line on the bottom of the piece until I get to the back of the gutter. Then I finish off the cut down the front of the gutter, eye-balling a plumb cut line. I finish the cut along the back with my tin snips. By the way, a slightly dull hacksaw blade binds less. Also, a loose blade will have a tendency to wander, so I make sure that the blade is very taut. I use a standard hacksaw, but I replaced the wing nut with a hex nut so I can really tighten the blade. It takes some practice to be able to make a cut without the blade binding, but as long as you are cutting straight and not applying too much pressure on the gutter, you

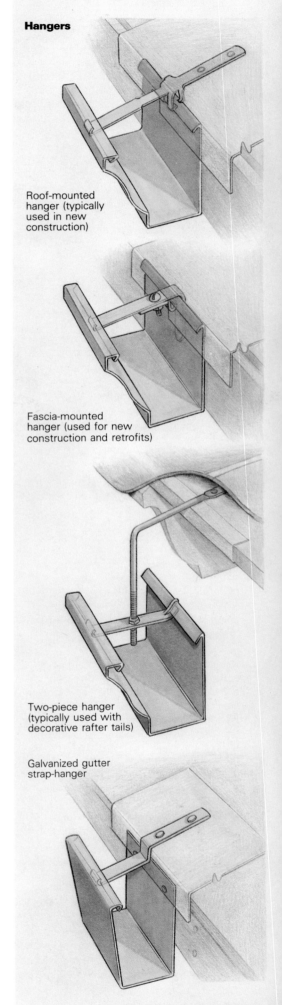

**Hangers**

Roof-mounted hanger (typically used in new construction)

Fascia-mounted hanger (used for new construction and retrofits)

Two-piece hanger (typically used with decorative rafter tails)

Galvanized gutter strap-hanger

should be able to accomplish the task without a lot of aggravation.

To install a miter strip, turn the gutter upside down, lining up the back of the miter strip with the back of the gutter. At this point you'll see how well you made your cut. If the blade drifted outward, you may need to trim the front with snips to make the strip fit well. If the blade drifted inward, you'll probably still be okay because the miter strip will cover the difference. Once the miter strip is lined up, drill a hole through the prepunched hole in the bottom of the strip and install a rivet; then do the other two.

**Rivet, rivet—**The secret to joining the modular elements in the aluminum gutter system is the pop rivet. Color-matched, aluminum rivets are available from the gutter supplier, and they can be installed on site in seconds using a hand riveter (photo right).

The miter strips and outlets are predrilled. Once the sections are in place, I use a small cordless drill with a ⅛-in. bit to make corresponding holes in the adjoining piece. Then I use a Duro Dyne PR-4 pop-riveter to install the rivets (Duro Dyne Corporation, 130 Rte. 110, Farmingdale, N. Y. 11735; 516-249-9000). This is the best hand-riveter I've ever used. It can accommodate several sizes of rivets, and replacement parts are available. I recommend getting extra springs because these will break occasionally.

A rivet joint isn't finished until it's been sealed with Gutterseal, a proprietary sealant supplied by the manufacturer. I don't seal the joints until the gutters are in place because the stuff can drip off a gutter that's being moved around, and it is very difficult to remove. When I seal a joint, I'm not stingy with the sealant (I'll typically use between one third and one half a tube per miter). All the rivets need to be covered, and I poke the tip into the joints to make sure I get full penetration. As it dries, the sealant shrinks significantly.

**Running gutter—**To hang the gutter, I first make sure that the nuts are loose on the hanger retainers. Then I hold the gutter in approximate position with its back parallel to the ground. I hook the front of the hangers into the front lip of the gutter and gently roll the back of the gutter upward into position, making sure I didn't miss any hangers before hooking the back lip into the retainers. I use a small pair of adjustable pliers to hold the nut as I tighten the retainers. If the shingles overhang the screw, this can be a tedious job that requires nimble fingers.

I continue to measure and install sections in as logical a sequence as possible, trying not to have two short sections coming together at a corner. It's best to install a piece of gutter with an outlet in it first; subsequent pieces should lap into it, allowing the gutter to drain completely. If a gutter section has miter strips on both ends, only one should be installed on the ground. The other should be fitted on the scaffold to make sure the inter-

**Mitering gutters**

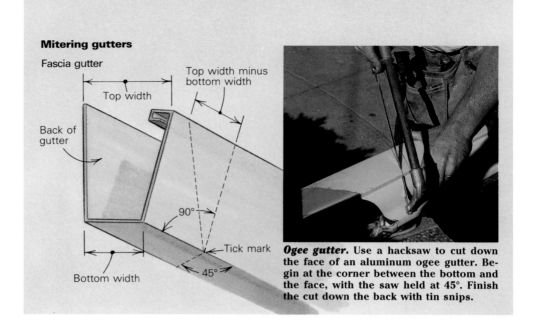

**Ogee gutter.** Use a hacksaw to cut down the face of an aluminum ogee gutter. Begin at the corner between the bottom and the face, with the saw held at 45°. Finish the cut down the back with tin snips.

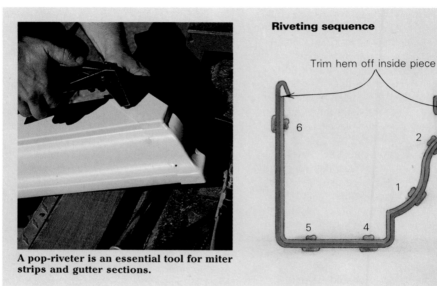

A pop-riveter is an essential tool for miter strips and gutter sections.

**Riveting sequence**

An expansion joint should be installed between sections of gutter more than 40 ft. long, or between fixed corners. Note how the hems have been trimmed off the piece on the left.

Drawings: Bob LaPointe

secting gutter doesn't end up pinched to its wall or hung out in space.

To splice two sections of gutter together, I cut 2 in. off the hem on the back and front lip so that it fits tightly when installed. Then I drill holes ½ in. from the edge of the outer piece—two on the bottom, three on the front and one on the back (bottom drawing, facing page).

On straight runs of gutter longer than 40 ft., or between fixed corners (as on a hip roof), aluminum gutters need expansion joints. Alcoa's version is a neoprene strip sandwiched between aluminum flanges that match the inside shape of the gutter. The flanges are riveted to each section of gutter, leaving the sections free to slide past one another as the gutters expand and contract. Straight runs of gutter are finished off with preformed end caps.

**Outlets, miters and downspouts—**I use an outlet as a template to mark its location on the bottom of the gutter. I draw a line on the inside of the outlet, and with an old wood chisel I punch a starting hole. Then, using my red-handled aviation snips (red for left cuts; green for right cuts; yellow for straight cuts), I cut just slightly wide of my line—just enough for the thickness of the metal of the outlet (photo, p. 127). I install the outlet through the inside of the gutter, tap it into place using a hammer and pop-rivet it.

Sometimes I have to make a section of gutter turn a corner that isn't 90°. Bay windows, which are typically 135°, are usually the reason. There aren't factory strips for these corners so I custom fit them, first bisecting the angle by making 67½° miters on corresponding sections of gutter. On one section I use my snips and hand seamers to make tabs that allow the sections to be riveted together (left photo, next page). They should be joined in place and liberally coated with sealant.

In addition to draining more water, larger downspouts are less likely to clog with leaves at the elbow where the downspout bends toward the house. The elbows and the section of downspout between the outlet and the wall is called a return. The piece of downspout between the elbows should be long enough to allow 1½ in. of overlap at both ends.

**Installing galvanized fascia gutters—**A fascia gutter is a no-frills, U-shaped gutter with a face that angles outward (top photo, p. 122). Fascia gutters come with or without wings—a combination nailing flange/flashing that extends up the back of the gutter to bear on the roof sheathing. On the plus side, the wing adds rigidity to the gutter and it's easy to install on a new roof before the shingles go down. On the minus side, it's hard to slope a gutter with a wing, and it's difficult to tuck one under existing shingles during a retrofit. So as with the aluminum gutters, I use an independent flashing under existing roofing.

Fascia gutters are anchored to the building by 8d galvanized nails driven through the back of the gutter into the rafter ends. A combination strap-hanger nailed to the roof sheathing

## Soldering galvanized sheet-metal

Soldering gutters is a difficult skill to master. I've soldered thousands of roof outlets on a bench, and I still find it tough to solder gutters. The vertical seams are the main problem because the molten solder has a tendency to drip off the iron rather than onto the work. But for those of you who think you can do it, here is how it should be done. It's a skill that's also handy for assembling custom galvanized flashings.

Successful soldering depends on several things, including a properly tinned and heated soldering iron (sometimes called a soldering copper), proper flux, the correct tip and properly prepared work. I typically use a 1-lb. iron for this type of work. You can get the tools and materials at a sheet-metal supply house.

Tinning the soldering iron is simply coating the faces of the tip with solder. To do this, heat the iron red-hot over a propane furnace and then clamp the iron in a vise. File all four facets of the tip with a coarse file, followed by a finer one. Next, melt a few drops of 50/50 all-purpose solid solder on a block of sal ammoniac (ammonium chloride) and rub the faces of the iron back and forth across the block until they are coated with a thin layer of solder. Do this outdoors, and don't breathe the toxic fumes that billow up during this operation. During soldering, I keep the iron as clean as possible with a wire brush. I will usually need to re-tin my iron after a day of constant soldering.

Now that your iron is properly tinned, you are ready to solder a joint. Start with a piece on a level surface. At the end of the seam, apply a small amount of pure muriatic acid with an acid swab (top left photo, below).

Wear eye protection, don't get acid on your skin and once again, avoid breathing the fumes. Place the iron at the end of the joint and melt a small amount of solder onto the metal (bottom left photo, below). Remove the iron, but keep pressure on the seam with the bar of solder until the solder cools a little. Keep the face of the iron on the upper piece of metal and try to sweat the solder between the joint by moving the tip back and forth across the joint. Don't try and do too much at one time—you may remelt already soldered areas. Do about ½ in. at a time, and then stop to let the solder cool slightly before continuing. Reapply the acid occasionally, if necessary.

The iron should be a little shy of 800° F for best results. If it's smoking or if the solder is turning black, it's too hot. There are lots of variables, but figure that you will have between one and two minutes of working time between reheatings.

Once you've completed a level seam (large photo below), try a vertical one, like the backside of a splice between two gutters. The trick is to apply as much solder as possible to a vertical joint, from the top down, without it all running or dripping to the bottom of the gutter. Once the back is complete, do the front and then the bottom using the solder that didn't stay in the joint. If your first attempt is anything like mine was, you probably will have enough excess solder to finish the rest of the job. Remember, the idea is to sweat the solder between the two pieces of metal, similar to sweat-soldering a copper plumbing fitting. The difference is that the solder is drawn into the joint with the heat of the iron and not a torch. If you try to solder gutters with a torch, you will only burn off the galvanized coating on the steel and the solder will never stick. —*L. W.*

Mitered angles that don't fit off-the-shelf miter strips are reinforced with tabs and rivets. Note how each tab has been formed by removing V-shaped sections on either side. At the gutter's contours, the point of the "V" corresponds to the inside and outside corners.

A mitered corner on a fascia gutter is attached by tabs on the end of one of the gutters. Here Williams uses a pair of hand seamers to bend back the solder tabs.

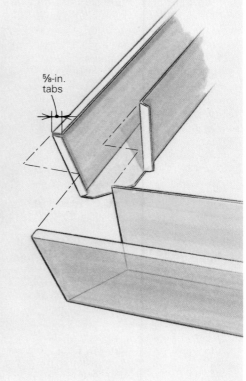

⅝-in. tabs

keeps the front of the gutter straight and braces the gutter against ladders.

I measure the gutter sections from corner to corner as with the aluminum gutters, but accuracy is a must because I don't have the luxury of the corner strips to make up for slack. Always take measurements from the back of the gutter, and lay out miters from those points.

**Outside miters**—Let's install a section with a typical outside miter. This section will have no tabs. Start by scribing a line with a scratch awl at 45° across the bottom of the gutter. Make a small tick mark where the line meets the face of the gutter, and square a line up to the top of the gutter (top drawing, p. 124). Next, measure over distance equal to the top width of the gutter minus the bottom width, and draw a line from there to the tick mark. Cut the gutter on that line with the appropriate snips. If you're cutting farther than 12 inches from the end of the section, cut the gutter outside the layout line with a hacksaw and clean up the cut with the snips (snip tip: when cutting, don't close the blades all the way—this will cause a burr like the barb on a fish hook).

Once the old gutters are removed, mark the rafter centers with a lumber crayon on the shingles. With a helper, hold the gutter in place and align the back of the miter with the corner and let it extend past the corner no more than ¹⁄₁₆ in. Hold the gutter as high and as straight as possible and tack an 8d nail at the high end of the gutter. Using a level, lower the other end to ensure a positive flow toward the outlet (½ in. in 20 ft. is plenty) and tack

another nail in place. If the piece is long, you may need to tack another nail at the middle of the gutter to keep it from sagging. If everything looks good, nail it off every 2 ft. to 4 ft.

Now slide the flashing under the roofing felt and over the sheathing until it is flush with the back of the gutter. Secure it with roofing nails. If nails from old roofs, flashings or hangers block the new flashing, pull them with a pry bar. A slater's bar can be a big help in yanking out the old nails.

To install a strap-hanger, hook it into the front of the gutter by pushing the gutter's lip inward a bit with your fingers until the strap's hook engages. Set the strap so its depth gauge rests against the back of the gutter and nail it through the sheathing into the rafter. Adjoining sections of gutter should overlap one another by 1 in. for rivets and soldering. Gutters without corners terminate at preformed gutter-ends.

Galvanized gutters also need expansion joints in straight runs of more than 60 ft., or in runs over 40 ft. between two fixed corners. I make an expansion joint by overlapping—but not connecting—two sections of gutter. I put a gutter-end flush at the end of one section of gutter. This section fits inside the adjacent section, which has a gutter-end recessed about 3½-in. from its end. This spacing allows a 1-in. gap between the two sections of gutter and a 2½-in. lap. It doesn't matter how the overlaps are ordered because both sections require their own outlets.

**Miter tabs**—The galvanized gutter that meets our mitered corner will have a miter cut running in the opposite direction, and it needs

soldering tabs for attaching the two pieces. They should be about ⅝ in. wide. Notch the tabs to measured layout lines and bend the back tab to 90° and the front tab to about 80° or 85° with hand seamers (top right photo).

Once I get the miters to fit properly on the wall, I join them with several steel pop rivets (three in front, two in back) prior to soldering them. The joints in galvanized gutters can be waterproofed with sealant, but the better method is to solder the joints (see sidebar, previous page). If you use a sealant, the best I know of is a polyurethane called Vulkem (Mameco International Corp., 4475 E. 175th St., Cleveland, Ohio 44128; 216-752-4400).

**Downspouts and returns**—Galvanized downspouts come in many standard sizes of round and rectangular shapes. Standard sections are 10 ft. long. Diameters of round pipe range from 2 in. to 6 in., in 1-in. increments. Rectangular downspouts range from 1¾ in. by 2¼ in. up to 3 in. by 4 in. The standard gauge for downspouts is 28 ga., but you can special-order heavier ones. Elbows are available for most downspout stock, but be sure to inspect them carefully for manufacturing defects such as holes at the corners or open seams—that's where they'll leak.

To make up a return using elbows, slip two elbows together as far as they will go, but don't force them. Hold the elbows on the outlet and measure the distance from the wall to the back of the elbow closest to the wall, on the angle of the elbows, and add 1½ in. Cut a piece to that length, but don't cut it off the

## Making custom returns

In addition to being more attractive than returns with elbows, custom returns can be used in situations where elbows take up too much space. In this example, the return is at a 45° angle to the wall, creating an outside angle of 135°. The 67½° layout line in the drawing to the right bisects the outside angle.

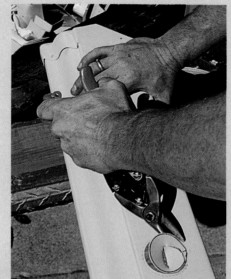

**After making an entry slit with a chisel, Williams uses a pair of aviation snips to make a cutout for a downspout outlet.**

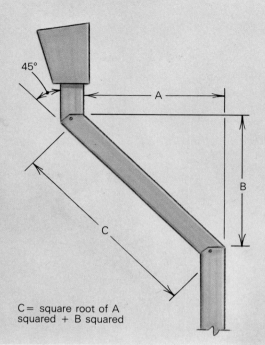

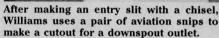
C = square root of A squared + B squared

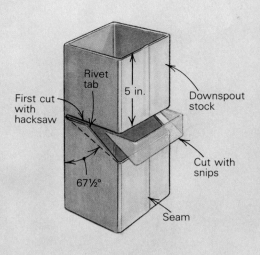

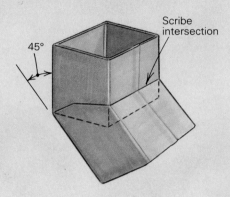

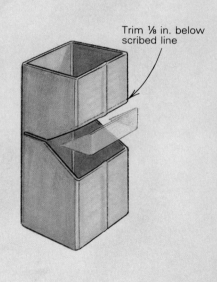

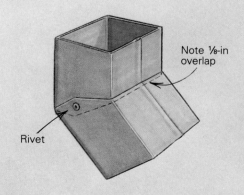

crimped end—there's a tool for that called a single-blade hand crimper made by Malco Products, Inc. (Highway 55 & County Rd. 136, P. O. Box 400, Annandale, Minn. 55302; 800-328-3530). Crimp one end of the piece, beginning with the four corners and then the four sides. Slip one elbow into the uncrimped end of the piece and then fit the crimped end into the other elbow. Remember that each piece slips into the next from the top down. Put the rivets on the top, or high up on the sides of the elbows, or they will leak. Never use 90° elbows for returns because a horizontal return will rust out prematurely.

At the wall, the return slips into a section of downspout. The downspout should be plumb and anchored to the wall at top and bottom with galvanized straps. Rivet the top elbow to the outlet. If the downspout is over 10 ft. long, keep adding full pieces and strapping them to the wall. I try to conceal the joints with the straps.

**Custom returns**—Sometimes elbows won't work for a return. For example, let's say that you have a gutter where the back of the outlet is 4 in. from the wall. You can't make the return with elbows because a couple of stock ones take up about 6 or 7 in., and to trim them is to ensure a leak.

To make a custom return, measure down about 5 in. on a piece of downspout, and scribe a line along the back—that's the seam side (drawings at right). Extend the lines along the sides of the downspout, and then mark a line from front to back at a 67½° angle. Use a

hacksaw to cut from back to front along the first cut line, but stop when you're finished cutting the sides. The front portion of the downspout will work as a hinge when you bend it. Use snips to cut the sides and back, leaving tabs on both sides for rivets. Now fold the top portion down, inside the tabs, and mark the trim line along the back as shown.

To calculate where the lower elbow occurs next to the wall, use the Pythagorean theorem (drawing above). Repeat the process to make this elbow, but make sure to fold the rivet tab inside at the bottom.

**Gutter care**—Gutters will load up with wind-blown dirt, leaves and the tiny aggregate that protects asphalt shingles. The extra weight puts a strain on the hangers and if the mix stays soggy, it will work to rust out galvanized gutters prematurely—especially if the gutters collect acetic debris such as redwood or pine needles. Keep them cleaned out. Galvanized gutters will last longer if they are painted on the inside. Prepare them first with a metal etch applied with a pump-type sprayer. Then paint them with latex primer and paint made especially for metal.

Outlets with leaf strainers—the ones that look like upside-down whisks—are less likely to clog when big leaves accumulate. From my experience, screens over the tops of the gutters don't do much good. □

Les Williams owns and operates L. A. Williams Gutters and Sheetmetal in Richmond, California. Photos by Charles Miller.